METAL BOATS

METAL BOATS

Ken Scott

SHERIDAN HOUSE

First published 1994 by
Sheridan House Inc.
145 Palisade Street
Dobbs Ferry, NY 10522

Library of Congress Cataloging-in-Publication Data

Scott, Ken (Ken H.)
        Metal boats / Ken Scott.
                p.   cm.
        Includes index.
        ISBN 0-924486-68-6
        1. Steel boats—Design and construction.   2. Aluminum boats—
Design and construction.   I. Title.
        VM321.S412   1994
        623.8' 182—dc20   94–20414
                                                                                  CIP

Photos by the author.
Drawings by the author unless otherwise indicated.

Printed in the United States of America

ISBN 0-924486-68-6

To my wife, Andrée,
fellow boatbuilder and sailing companion

Contents

Foreword

Steel or aluminum will give you many advantages over other boatbuilding materials. Problems such as delamination and gel coat bubbles will soon fade from your memory. Those hard-to-find leaks, where the water seeps in under a stanchion or at the hull-to-deck joint and then travels behind the inner lining to reappear at a distant point, will no longer interest you. A fiberglass boat usually consists of a hull and deck moldings that are stuck together, then many holes are drilled in the deck to attach stanchions, winches, blocks, etc. On a metal boat, all of these items are welded together to form a really watertight shell.

The hull of a metal boat will usually cost less than a custom-built fiberglass hull, even though the mast, rigging, sails, engine, and electronics cost the same whatever the hull material. Not only that, it will be much stronger and can reflect your own personality more than any mass-produced boat could.

When a fiberglass boat hits a rock with her keel, the hull may crack, the keel bolts may loosen, or the keel may even be ripped off. A metal boat will usually collect a dent in her keel that can be filled at your leisure during the next haulout. And when you drag ashore in that isolated anchorage....

After serving 26 years in the Canadian Navy, I knew firsthand the advantages of steel as a shipbuilding material. But when I started on my boat, I could find very little published on small metal boats. I wrote this book to fill that information gap and to provide you with the fundamental skills and knowledge required to build or buy a steel or aluminum boat. It is also a photographic chronicle of the work done on *Scot-free II,* the boat shown on the cover, as she was built and fitted out. Where I had to imagine a detail or picture in my mind's eye, you will be able to see it in clear black-and-white photographs.

Ten years and 11,000 miles later, I am still happy with my steel boat. I hope this book will help you to build or buy a better boat.

Ken Scott
Victoria, British Columbia
June 1994

METAL BOATS

A big steel boat.—The American aircraft carrier
USS John F. Kennedy, *displacement 85,000 tons,*
at Norfolk, Virginia. I wonder why they didn't
build her out of wood, glass-reinforced plastic
(GRP), or cement!

CHAPTER

1

The Improbable Dream

Most people who consider metal boats have some experience with wood or fiberglass boats, and they want to know the advantages and disadvantages of buying or building a boat in steel or aluminum. Although a boat buyer doesn't need the skills of a boatbuilder, he or she should still have enough knowledge about building techniques and problems to know if the boat being considered is well built. Boats are expensive, and a study of how they are made might be the best investment ever made by a buyer/builder. So although this book is written from the viewpoint of an amateur boatbuilder, it is meant for anyone building, finishing, or buying a metal boat.

The urge to build a boat or to finish a bare hull hits people in different ways. You might have owned several boats and never been able to find one that just suited you; you might have seen an old boat and fallen in love with it and thought, "If only the same thing were available in modern materials or with modern lines or rigging." Or perhaps you want a big boat to sail offshore but can only afford a small one, and decide that the only way you can get the boat you want is to build it yourself. No matter how you got to this point, as soon as you decide to build your own boat, you will find that you are drifting away from the majority of your friends and neighbors into a private world, shared only by a few odd people with whom you can discuss your ideas and problems.

At the time this mental aberration finally took firm hold of me, I had had about 25 years experience as a sailor, naval officer, and marine engineer. After 25 years, I knew almost exactly what I wanted in a boat; my problem was to get as close to that ideal as possible at a price I could afford. Still the last few months before taking the plunge were full of doubts.

Because my interest has been mainly in sailing, this book is biased toward sailboats, but most of the material is equally applicable to powerboats.

HOW BIG SHOULD OUR BOAT BE?

From the beginning you will know whether your dream boat is to be sail or power, and your first decision is how big you want it to be. "As big as possible," you say? Then a realistic assessment of the crew you will have available is a must. Large boats require large crews, especially sailboats. So, although it's nice to have a huge boat on which to take all your friends out for an occasional weekend, think about who will be available for extended cruises—you will most likely conclude that it is just yourself and your spouse or friend. For weekend races, it is difficult to get a crew together and keep it together for a season; for a long cruise, it is practically impossible. And being dependent on other people takes much of the pleasure out of the freedom found in cruising. "The kids," you say? Yes, but those eager teenagers will have grown up by the time you finish the boat, and will be more interested in their dates than in their parents' activities.

Larger boats also require more money and time to build and maintain. Fittings are more expensive. Sails are priced by the square foot and boats can be costed by the pound. Every shackle, line, valve, etc., will cost more as the boat gets bigger. Comparing two sailboats by the same manufacturer, of the same basic design, we have:

Length	35 ft	41 ft
L.W.L.	30 ft	35 ft
Beam	11 ft, 6 in.	13 ft, 4 in.
Draft	4 ft, 6 in.	5 ft, 6 in.
Displacement	13,800 lbs	25,600 lbs
Sail area	480 sq ft	670 sq ft

If you estimate the cost of a boat by its displacement (which is really how a boat is priced, not by the foot), you will see that, for the extra six feet of length, the price of the boat will practically double. The cost of all sails will be much higher, a much larger engine will be required, therefore, more space for fuel will be needed, and the list goes on.

The prices of a few items follow to illustrate how small increases in size can cause large increases in cost. The prices are given in American dollars in 1993, but the comparison would be valid in any currency.

Engines:

27 HP	$ 6,600
44 HP	$ 9,900
66 HP	$11,000

Braided line:

⅜ in.	$0.37/ft
½ in.	$0.60/ft
⅝ in.	$0.94/ft

Bruce anchors:

22 lb (10 kg)	$310
44 lb (20 kg)	$564
66 lb (30 kg)	$820

In addition to the cost and work involved with building a larger boat, think of the cost and work to keep it in good shape year after year. The bigger the boat, the more time you will spend taking care of it and the less time you will spend enjoying yourself. Haul-outs will cost more for a larger boat, and in many small yards, facilities to haul out large boats will not be available. The cost per foot when you stay at a marina is also a consideration. A boat is rather like a house: if you buy one that is too big, you spend all your time looking after it rather than just enjoying it.

There is also the question of the physical strength required to handle a large boat. Although this is not as applicable to powerboats, sailboats are basically manual machines. Have you ever tried stepping a 50-foot mast? It is a daunting project. Though a 40-foot mast can be handled by a couple fairly easily, a 50-footer requires extra hands and some careful planning. The forces generated by large sails have to be experienced to be believed. While a racing crew of three men on the foredeck might be able to change a large headsail with no problem, when you try to change the genoa by yourself while your mate handles the helm and sheets, you will find it physically challenging. Even if you can handle a big boat now, will you still be able to handle it in ten years? Or if you suffer an injury at sea and can only use one arm, will you be able to handle the boat?

And what about big powerboats? That huge powerboat might look very nice alongside the floats but think of it alongside the fuel dock as you feed its thirsty engines. Big boats = big engines = big fuel bills.

For the most part, cruising crews consist of two people, usually a couple that's been together long enough to be able to put up with one another in confined quarters. A reasonably sized boat would provide comfort for two people full time, and adequate accommodation for a couple of visitors for a short time.

So, what is a good size for a cruising sailboat? Anything under 30 feet makes it difficult to carry enough food, water, tools, and equipment for comfort and safety. You will be living on this boat full time, not just for a couple of weeks in the summer, so you will want to carry with you many of the things you use at home, such as a radio, books, cassettes, fishing gear, diving gear, warm-weather and cold-weather clothes, tools, and spare parts. I would suggest that 34 to 40 feet would be satisfactory for most people.

Should it have a center cockpit or aft cockpit? The aft cockpit has the advantage of being farther away from the bow spray than the center cockpit, and of keeping the internal space as one large area; also, it is usually lower at the stern, making it easier to enter from a dinghy. A center cockpit boat will make it easier to mount the engine closer to the center of gravity of the boat, and in a location that will give easier access to it—no need to try and push it as far aft as possible, which is usually the case in an aft-cockpit boat. It also has the advantage of offering two separate living areas below deck, including a private sleeping cabin aft. This is convenient if you normally carry children aboard or if two couples are sailing together. This is the arrangement normally found on charter boats. But don't forget, charter companies hope to charge as much as possible for their boats, and if they make it convenient for more people to use the boat together, it will reduce the cost per person. Not the same aim as the person building a boat for personal use. The disadvantage of a center cockpit, of course, is that there are two small below-deck spaces rather than one large one. That is not too bad if you are going to charter for just a couple of weeks, but it can get rather claustrophobic if you are living on your boat full time. The smallest boat in which it is reasonable to have a center cockpit is about 42 feet. The bigger the boat, the better this arrangement works.

One mast or two? For anything under 38 feet, the extra windage of the second mast hardly seems worthwhile. By using a cutter rig, you can keep the size of each sail within the capability of the average couple to handle. If you use roller furling on the jib, then you can reduce sail without having to go forward of the mast.

For a boat 40 feet and up, two masts in a ketch or schooner arrangement become more appropriate. The combination of sails that can be used will delight the keenest sail changer. Add a main and mizzen staysail and a main and mizzen spinnaker, and you can entertain yourself on the calmest day. You may not go any faster, but you'll get lots of exercise. A yawl with its tiny sail aft doesn't seem to be an efficient rig, because the mizzen is too small to make the extra windage of the second mast worthwhile.

STEEL OR ALUMINUM?

Once you have decided on the size and type of boat you want, the next question is steel or aluminum? Steel has these advantages:

1. It is widely used; therefore, many yards are capable of working in this material, and it will be easy to have damages repaired.
2. It is easy to weld.
3. It is inexpensive and strong.

Its disadvantages are:

1. It is subject to corrosion. Steel boats need expensive surface preparation and painting to protect them against rust.
2. It requires upkeep. The battle against rust is an ongoing problem and chips and scratches in the paint system must be constantly repaired.
3. It is harder to cut and shape than aluminum.
4. The panels are heavier and so are a bit more difficult to handle than aluminum when building a boat.

Aluminum has these advantages:

1. It is corrosion resistant.
2. It is lighter in weight, which is an advantage both in building the boat and in its final weight.

3. Due to its light weight, many interior parts can be made out of aluminum that on a steel boat would be made out of wood—hatches, furniture and cabinet frames, etc.

4. It is easier to cut and bend.

Its disadvantages are:

1. It is more expensive.

2. It is more difficult to weld.

3. It is more susceptible to electrolysis.

4. It is difficult to find antifouling paint now that some governments have restricted the use of tin-based antifouling paints. (Most antifouling paints contain copper, which will damage an aluminum hull.)

BUILDING THE BOAT

The next important decision is how much to do yourself. One extreme is to do everything yourself: choose your trees, fell them, saw the logs into planks and dry them or, for a metal hull, smelt your own ore and roll the metal. The other extreme might be just to make the winch covers and curtains. Between these two extremes lie an infinite number of possibilities. Even when you've decided on your major starting point, you will still have a great number of small choices to make. For example, are you going to make your own cleats and stanchions, or buy them and weld them to the deck? Each choice will result in a different allocation of time and money to be spent on the boat.

First, are you going to build the hull yourself or have it built? If you build the hull yourself and hold down a full-time job at the same time, you can expect the project to stretch out eight or ten years into the future. Do you want to give up that much of your spare time, or would you rather go sailing and have a more normal social life? If you have the hull built and finish it yourself, the project will diminish to a more reasonable three to five years of your spare time. Still a long period, but at least you'll be done in the foreseeable future. A professional boatbuilder can produce in four or five months a hull that would take the amateur boatbuilder several years of spare time. While the hull is being built, you can help the project along by making items for the boat, accumulating lead for the keel, and shopping for parts and material.

One of the many advantages of a metal hull is that you can take a standard design and have it modified to your taste at very little extra cost. Since there is no mold needed, as with a fiberglass boat, the builder can easily make changes to a design to suit the owner. So you can have a custom design in metal more cheaply than a custom-made fiberglass boat.

The cost of having a metal hull built, especially a hard-chined vessel, which can be built in a small yard without expensive machinery to roll and shape the metal, is not much more than the cost of building it yourself. If you build the hull yourself, the first thing you will need is a place to work that gives shelter from the weather. This will entail either renting a large building or constructing a temporary shelter. An electrical supply will be needed and, for northern climes, a good heating system. The work will be dirty and noisy, so finding a place to work may prove more difficult than you expect. Many yacht clubs and marinas will not allow this type of work to be done on their grounds, and most community by-laws or neighbors will not allow it either. For most people, the savings are not worth the time and effort.

On the other hand, if you have the hull built, it can then be put in your backyard or almost any convenient place while you finish the inside. The finishing work is fairly clean and mostly inside the boat, so the neighbors will not complain, and no shed is needed because the boat itself is weatherproof as soon as the hatches are installed.

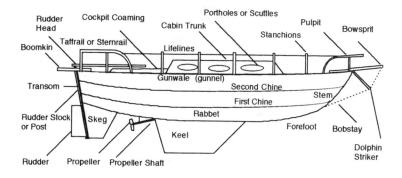

Figure 1.1 Parts of a boat (profile).

The discounts available to boatbuilding yards that are not available to the private builder are another financial consideration. A yard will be able to buy an engine and install it for about the same price as an individual would pay for the engine alone, and the metal for the hull will probably cost you retail at the supplier's shed, while the commercial builder will be paying wholesale prices delivered to the boatyard.

MATERIALS AND TIME

Before starting to build a hull, you should obtain an estimate from a builder for constructing the same hull, then make a careful estimate of your costs to produce it. A checklist follows that will help you estimate your hull-building costs.

Shed, rent or build
Electricity
Water
Heat

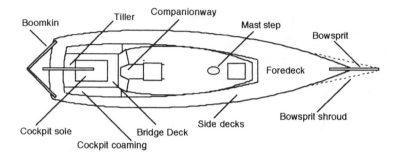

Figure 1.2   Parts of a boat (plan).

Tools: welder, nibbler, drill press, large grinder,
 oxy-acetylene torch (to heat and bend metal),
 electric welder

Consumables: Welding rods, drill bits, grinding
 disks, etc.

Steel or aluminum material: sheet metal, stringers,
 frames, construction jig, etc.

Crane to turn hull over

Sandblasting (for steel hulls): cost of sandblasting
 and cost of grit (which is surprisingly expensive)

Paint: Steel needs to be primed with an expensive
 zinc coating; epoxy undercoating paint, polyurethane finishing paint

You'll also need to put aside time for the following tasks:

Learning how to weld

Finding a building

Buying machinery

Buying materials

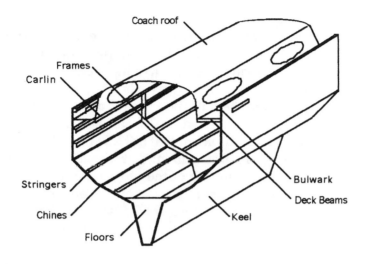

Figure 1.3 Structural terms.

Drawing layout plans in full size
Building the construction jig
Building frames
Building the hull
Sandblasting (if steel)
Traveling time to and from the site

A third alternative to either building or buying a hull is to build the construction jig, cut out and weld together the frames and stringers, and tack-weld the plating. Then hire a professional welder to weld the seemingly miles of difficult welds on the thin sheet-metal plates. This will allow you to do the time-consuming metalwork and the simpler welding, and leave the more difficult welding for someone who knows the trade. It will also ensure good-quality watertight welds. In other words, a better hull in less time.

SUMMARY

Building a hull is a huge job and if I seem to be discouraging you, I'm doing so on purpose. While shopping for parts for my boat, I answered an advertisement in a newspaper and met a man who had spent ten years working on a concrete boat. He had finished the hull and the deck and had bought many parts, but the project had finally defeated him and he was sadly selling the equipment that he had bought. During our conversation, it became evident that he had had a wife and children at the beginning of the project, but there was no sign of them now. On another occasion, a woman I spoke to, while shopping for three tons of lead, said that her husband had spent years saving lead but had died of a heart attack before starting his boat building project.

The morals of these stories are:

1. Look for ways to shorten the time required for the project by carefully assessing what you should do yourself and what you should buy or pay someone else to do.

2. Don't wait until you are ready to retire before you start, because boatbuilding is hard work and if you wait too long you may not live to enjoy your ship.

On a more upbeat note, one gentleman I met had started building a hull from scratch when he was 65; it took two 6-month full-time periods to complete it, and he had just returned from a long cruise. Of course, he was a metalworker by trade, so that helped speed up the first half of the project, and the inside of his boat was rather simply finished—but he had done it!

Victoria Clipper IV

This modern aluminum catamaran is in service as a ferry between Victoria, British Columbia, and Seattle, Washington. There are four of these vessels—each slightly different because the design is improved with each hull. They were built by Fjellstrand in Norway. The local waters are strewn with logs from broken log booms so these vessels use waterjet propulsion rather than conventional propellers.

Length	38.8 m
Beam	9.4 m
Draft	1.5 m
Tonnage	420 GRT
Power	Two 2010 BHP engines
Speed	31 knots
Passengers	300

CHAPTER

2

All Shapes and Sizes

One of the many advantages of metal hulls is the ease with which changes to a basic design can be made. It is possible to take a well-proven design and alter it to your own needs. If this is done before the hull is constructed, there is little extra cost. Even after the hull is finished, if you do not like the results, you can always cut out the offending bit and weld another in its place. The resulting boat will be just as strong as the original. This type of modification was often seen in the aluminum 12-meter boats that used to be sailed in the America's Cup races. The designers often changed their underwater shapes to try to squeeze that extra tenth of a knot out of their masterpieces.

However, any major change should be examined by the naval architect who designed the boat to ensure that it does not make the boat less efficient or even dangerous. Also remember that some day even the best loved dream boat will be sold, so if you make it too unusual it will be difficult to sell.

Another advantage of a metal boat is that, should it be damaged at some time, it will be relatively easy to cut out the damaged piece and weld in a new one. Also, after an unexpected storm hits a remote anchorage you'll often see fiberglass boats with holes in their sides lying next to metal boats with dents. The dents aren't pretty, but the boats still float!

We did some practical research into this a few years ago. While we were ashore in the Dominican Republic, a boat dragged down on ours and pulled our anchor out. The two boats finished up ashore pounding against a beached ship, with our boat in the middle, like the meat in a salami sandwich. When we returned we shoveled bucketfuls of fiberglass off our decks, but our hull was still watertight. The large dent in the starboard side gives *Scot-Free II* that hand-crafted look. One day we must get it pushed out.

Figures 2.1 and 2.2 show two variations on a Bruce Roberts 36. Figure 2.1 shows the original design with its full-length keel, very curved sheerline, low headroom, and short cabin.

Figure 2.2 shows the author's variation on the same theme. The keel has been shortened and slightly deepened to increase the boat's maneuverability in confined waters. It has been fashioned in the shape of an airfoil cross section to decrease the drag and improve the boat's performance. A stern-hung rudder has been added. (A stern-hung rudder is further from the turning point of the boat, so less force needs to be exerted by the rudder to turn the boat, therefore causing less drag.) A large skeg forward of the rudder protects it and provides directional stability. The stern-hung rudder also makes it easy to add a self-steering wind vane, using a control tab on the trailing edge of the rudder. If the rudder is damaged, this location makes it easily removable for repairs. This arrangement also saves you one hole in the hull for the rudder shaft.

The sheerline has been raised by six inches amidships to increase the reserve buoyancy and to provide an extra six inches of headroom inside. The cabin has been extended for-

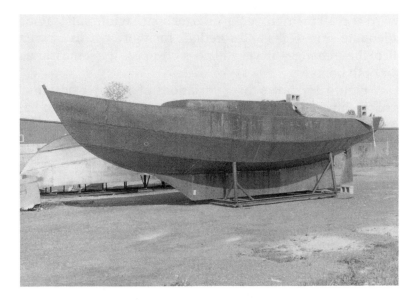

Figure 2.1 A Bruce Roberts 36 in steel.

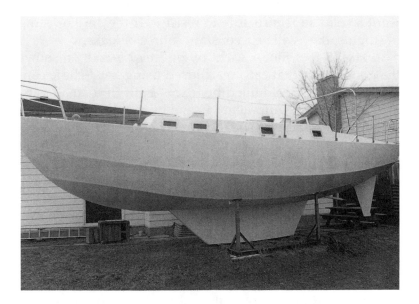

Figure 2.2 Author's variation on the Bruce Roberts 36 of Figure 2.1.

ward and aft to improve the interior space while still leaving adequate side decks and foredeck for working. Even with all of these changes, it was still possible to use the same frames for both boats.

Figure 2.3 shows three variations of an older design of a hard-chine, shallow draft, steel-hulled boat. One owner has chosen a rear cockpit, two have chosen center cockpits. The middle boat has a large aft cabin, with a raised stern and portlights built into the hull aft. This has created a large state-room aft with plenty of light and air.

The owner of the boat in Figure 2.4 (at the far left in Figure 2.3) has chosen a ketch-rigged, aft-cockpit version. When photographed, the boat had been coated with zinc primer only and had not yet been painted. The inside was not finished. The owner decided to go sailing before he fin-ished the inside of the boat—not a good idea, because once you start sailing you will never get around to putting in the hundreds of hours needed for the detail work inside. Hav-ing portlights on the forward end of the cabin is a weak point in this design: when solid water comes over the bow, you want a solid superstructure to resist it. If they are opening portlights, they will be a constant source of leaks.

The 55-foot motor yacht of Figures 2.5 and 2.6 shows the good shape that can be achieved using hard-chine con-struction. This steel-hulled boat was built in a small yard using no special equipment to bend or roll the metal. It is a credit to the good plate development carried out by the de-signer.

In planning the general appearance of a boat, one should ensure that all sloping lines are at the same angle. For ex-ample, the coamings above and below the main steering po-sition (Figure 2.6) have the same slope on their forward faces; the deck supports have the same forward slope and the stan-chions should either match this slope or be vertical.

With this type of vessel, if the hull is built of steel, con-sideration should be given to constructing the superstruc-ture of aluminum. This will save hundreds of pounds of topweight, giving greater stability and better fuel consump-

Figure 2.3 Three variations on an older design.

Figure 2.4 A classic ketch.

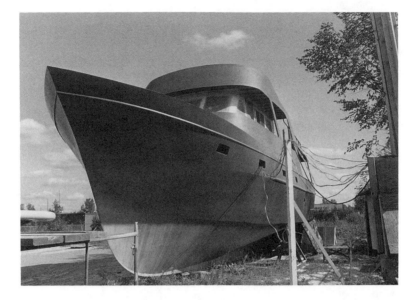

Figure 2.5   A 55-foot motor yacht.

Figure 2.6   Motor yacht.

tion because the boat will be lighter. If this mixture of metals is used, the two metals must be insulated from one another to prevent electrolysis. This can be done by riveting or bolting the two metals together with an insulating material between them, an insulating sleeve on the rivet or bolt, and an insulating washer between the rivet head and the opposite metal (that is, if you are using aluminum rivets, put the washer between the rivet head and the steel). Another system uses a specially manufactured bimetal strip with steel on one side and aluminum on the other. These two metals are explosively joined to make a permanent bond. The steel hull is then welded to one side and the aluminum superstructure is welded to the other, forming a corrosion-resistant joint.

Aluminum superstructures on steel hulls have been used on Canadian naval ships for the past 40 years, so this is not a new technique. If you decide to change an existing design, have the naval architect check the stability calculations to allow for the reduced topweight.

A Pedro 1000 is shown in Figures 2.7 and 2.8. This 10-meter boat was built in Holland at Pedro Boot BV, Zuidbroek, for a Canadian owner. She is of single hard-chine construction with no internal frames. The stanchions and lifelines are of welded steel pipe. Large windows will let in a lot of light and, one hopes, will also be strong enough to prevent the water from following the light. The hole in the stem, about half a meter below the deck, is a hawse pipe for the bow anchor, which slopes up to the deck-mounted windlass. This arrangement works well for a standard pattern stockless anchor, but is not as satisfactory for a plow anchor, because its blade chews up the paint on the stem. The internal view looking forward in the same boat shows a minimum number of stringers (Figure 2.8). The short reinforcing frames on the flat planing section of the bottom will prevent flexing in this area. At the lower left, the galley area has been framed in to try it for size. It is much easier and cheaper to use a few battens to help visualize the location of the furniture than to build it and then find out that

Figure 2.7 A Pedro 1000.

Figure 2.8 Interior view of Pedro 1000.

Figure 2.9 Aluminum work barge.

things don't fit. The control station with its hydraulic steering mechanism is shown at the right.

The two great advantages of aluminum over steel are its light weight and immunity to rust. The work barge shown in Figure 2.9 is exposed to all kinds of weather and rough usage, yet it never looks shabby or needs painting. The time used to prepare and paint other materials can be used as working time on this vessel. The simple and effective design is appropriate for this type of workboat. It is similar to a military landing craft in that the front can be lowered and vehicles and equipment can be easily loaded. The outboard engine at the stern provides good maneuverability and a clear working deck forward.

Figure 2.10   Aluminum sailboat.

In contrast to the working barge is the well-built sloop of Figure 2.10. The metal has been rolled so skillfully that after it is painted it will be difficult to see any difference between it and a fiberglass boat. Rolling and fitting metal this well takes an experienced boatbuilder—not your average backyard amateur.

Pathfinder *plies the waters of Lake Ontario*

This steel-hulled brigantine is used as a training vessel for young people who want to experience the adventure of sail. Some of her vital statistics are:

Overall length	72 ft	Registered tonnage	31.6 tons
Length on deck	60 ft	Ballast	10 tons
Waterline	45 ft	Hull thickness	¼ in.
Draft	7 ft, 10 in.	Engine	110 HP
Beam	15 ft, 3 in.	Diesel fuel	400 gal
Mainmast height	48 ft	Sail area	2,450 sq ft

3

Materials

Many different materials are used in boat construction and the composition of each material is described in some sort of standard or specification. Most countries have their own standards association, such as the ASA (American Standards Association) in the United States, the CSA (Canadian Standards Association) in Canada, the BS (British Standards Institute) in Great Britain, and DIN (Deutsche Industrie Norm) in Germany. International standards are established by the ISO (International Organization for Standardization), which has its headquarters in Geneva, Switzerland. Scientific and most engineering work is carried out using the International System of Units (SI—Système International). This allows scientists and engineers from different countries to compare their findings and to communicate in a common technical and scientific language.

The ISO standards are often taken from those established by national organizations. For example, the International

System of Units is based on the French metric system, which
dates back to 1791. This in turn dates back to the seventeenth
century, when various groups, including navigators, sug-
gested that the basic unit of length measurement should be
the arc of one minute of a great circle of the earth. Photogra-
phers who have been used to seeing their film rated as ASA
200 will now find that it is ISO 200, because the ISO has
adopted the U.S. system for describing film speed.

Some countries still use the British Imperial system of
feet, inches, and pounds, from which we get fascinating tables
such as:

16 drams	=	1 ounce
16 oz	=	1 lb
14 lbs	=	1 stone
8 stone	=	1 hundredweight (cwt)
20 cwt	=	1 ton = 2240 lbs

Other countries have further complicated this system by using their own variations; for example, a British gallon of fresh
water weighs 10 pounds, but an American one weighs 8.3267
pounds. This makes your tankage calculations more complicated since, instead of a 20-gallon tank holding 200 pounds
of water, it holds 166.534 pounds in the United States.

Two sets of standards of particular interest in North
America are those of the ASTM (American Society for Testing and Materials) and the AISI (American Iron and Steel
Institute). The ASTM, founded in 1896, produces an annual
publication consisting of some 66 volumes covering standard
methods of testing and specifications for almost every material known to mankind—not the type of book to be found in
the average home library, but it should be available in a good
public library. The six volumes in the first section of the publication cover iron and steel products, the five volumes of
the second section cover nonferrous metals, and the four
volumes of section eight cover plastics. These are the volumes of most interest to the boatbuilder who would like to
know more about the materials of construction. For steel used

in boatbuilding, the AISI numbers are the designations commonly used.

STEEL

Steel is produced from pig iron by the removal of impurities and the addition of carefully controlled amounts of additives. If the major properties of the steel are a result of its carbon content, it is commonly called *carbon steel*, and this is the type used for general construction purposes and boatbuilding. When the amount of carbon becomes high (between 2% and 4%), the product is termed *cast iron*. When the steel's properties are due to some material other than carbon, it is termed *alloy steel*. This includes the stainless steels. Common alloying metals are nickel, chromium, molybdenum, and manganese.

Table 3.1   Some Typical Properties of Carbon Steels

AISI Type	1020	1040
Carbon content (%)	0.18–0.23	0.37–0.44
Tensile strength (psi)	75,000	97,000
Yield strength (psi)	64,000	82,000

As Table 3.1 shows, metal has the endearing property of getting stronger after it is stressed to its yield point.

STAINLESS STEEL

Stainless steels can be divided into two main groups using the AISI classification system. The 200–300 series are chrome-nickel alloys with an austenitic grain structure, and the 400 series are chrome alloys with ferritic or martensitic grain structures. (More information on the structure of metals can be found in books on metallurgy, but is not really necessary for the average boatbuilder.) Each material has properties appropriate for certain uses. For example, knife blades must be rust resistant and hardenable but need not be nonmagnetic. In the 400 series, the martensitic group is hardenable and magnetic, and the ferritic group is nonhardenable and magnetic. Therefore, the 400 series has an alloy suitable for knife blades. The 200–300 series stainless steels are tough, ductile, nonmagnetic and weld easily. So you will find that a magnet will stick to a stainless steel knife blade but not to the stainless steel normally used on boats.

The stainless steels commonly used in boatbuilding are alloys of chromium and nickel with iron, the 200–300 series (Table 3.2). By varying the amount of each component and the addition of other metals, the resistance to corrosion (and the cost) can vary. Extra low carbon content stainless steels are also available. These are indicated by the letter *L* after the number, for example, 304L and 316L, and they contain less than 0.03% carbon. These are particularly useful for welding rods, because the low carbon content reduces the corrosion and brittleness at the edge of the welds.

Table 3.2   Some Typical Properties of Stainless Steels

AISI Type	302	304	316
Composition (%):			
Chromium	17–19	18–20	16–18
Nickel	8–10	8–12	10–14
Molybdenum	—	—	2–3
Manganese	2.00 max	2.00 max	2.00 max
Carbon	0.15 max	0.08 max	0.08 max
Sulfur	0.03 max	0.03 max	0.03 max
Silicon	1.00 max	1.00 max	1.00 max
Corrosion resistance in marine environment	Good	Better	Best
Weldability	Good	Good	Good

ALUMINUM

Aluminum is the second most common metal in the world. In its natural state it is known as bauxite. This is refined to alumina (aluminum oxide), which is then, using a procedure of electrolytic reduction developed in France in 1886, reduced to pure aluminum.

Aluminum is characterized by its low weight compared to steel, its high strength (when alloyed), and its resistance to corrosion. Aluminum's corrosion resistance is due to the fact that when its surface oxidizes, the oxide forms an air-tight surface that prevents oxygen from reaching the base metal and continuing the reaction. This means that as soon as the whole surface oxidizes, the action stops. On steel, the iron oxide (rust) allows oxygen to permeate to the metal below such that the metal continues to oxidize. Hydrochloric acid will dissolve this oxide coating, so be careful not to spill your battery acid on your aluminum boat.

Pure aluminum is a very soft material but the addition of metals such as copper, silicon, magnesium, and manganese can produce almost any required property, such as the strength of steel, corrosion resistance, machinability, and weldability.

Many different standards are used to describe aluminum alloys; one that is commonly used in the United States is that developed by the Aluminum Association. This system describes wrought alloys with a four-digit number and casting alloys with a three-digit plus decimal point number. The aluminum wrought alloys have these designations:

Designation	Main Alloying Metal
1xxx	Pure aluminum
2xxx	Copper
3xxx	Manganese
4xxx	Silicon
5xxx	Magnesium

6xxx	Magnesium and silicon (magnesium silicide)
7xxx	Zinc
8xxx	Tin and lithium

Within these general groupings, some 260 types of alloys are available, with all types of properties such as strength, hardness, and corrosion resistance. In any marine application, the builder must be careful to use only the marine-grade aluminum specified by the naval architect. It has been chosen for its resistance to corrosion in seawater, its strength, and its ability to be welded. The designer has (we hope) spent a lot of time studying the various alloys available, and has consulted with the manufacturers for their recommendations. If the specified aluminum is not available, the builder should ask the designer's advice on a substitute material *before* taking some well-intentioned supplier's advice on aluminum that is "almost the same." The 5000 series is commonly used in marine applications, but within this group a wide variety of properties is still available.

Table 3.3 shows the variation in strengths of a few alloys; their other properties, such as corrosion resistance, will vary just as much.

Table 3.3   Some Typical Properties of Aluminum

Type	Tensile Strength (psi)	Yield Strength (psi)
2011	55,000	43,000
3003	16,000	6,500
4032	55,000	46,000
5005	18,000	6,000
5052	28,000	13,000
6010	43,000	27,000
7005	57,000	50,000

Aluminum parts often have their corrosion resistance increased by the process of *anodization*. In this treatment, the aluminum part is made the anode in an electrolytic solution and a surface of aluminum oxide is plated on; this is the same as normal oxidization only thicker and tougher. A thinner surface protection can be put on by simply dipping the aluminum part in an oxidizing solution.

BRASS AND BRONZE

In general, brasses are combinations of copper and zinc, and bronzes are combinations of copper and tin. In each case, small amounts of other metals are added to provide a variety of properties such as increased hardness, flexibility, corrosion resistance, etc. Lead, for example, is introduced into the alloy to increase its machinability.

Brass is attractive and makes very nautical-looking lamps for your boat. Polishing such lamps will give you something to do on rainy days as brass tarnishes quickly in the marine environment. Brass, however, cannot be used in saltwater systems because it suffers from the unpleasant problem of dezincification. When brass is exposed to a diet of saltwater, an electrolytic cell is set up, and the zinc is eaten away, leaving a weak porous structure of copper. This can be seen as the color changes from brass to copper. Brass is more decorative than useful in a marine environment.

Bronze is a combination of copper and tin, with other metals in small amounts—hence the importance of tin mines to the Roman armies for their swords and weapons and to the sailing navies for their bronze cannons. The bronzes have much better corrosion resistance than the brasses and are normally stronger. The main component of a bronze alloy is copper, the second is tin (from 2% to 20%), and other metals are added to give specific properties. A bronze will often take its name from this third metal. Phosphor bronze, for example, contains 4% to 6% tin and less than 0.5% phosphorus. This alloy has good resilience, is nonmagnetic, and does not corrode in a marine environment. Some "bronzes" are alloys of copper and some other metal: aluminum bronze consists of 88% copper, 8% aluminum, 3% iron, and only 0.5% tin, and yet it is normally referred to as a bronze.

MONEL METAL

This alloy is made up of 67% nickel and 28% copper. It is used for machine parts that need to be strong and corrosion resistant, such as seawater pump shafts. One of the variations of this alloy is *K-monel*, which contains aluminum and titanium as well as nickel and copper. This strong, corrosion-resistant alloy is nonmagnetic and can be used for special purposes such as main shafting on minesweepers or near magnetic compasses on boats and aircraft.

Table 3.4   Weights of Material

Material	gm/cm^3	lb/ft^3
Aluminum	2.7	165
Brass	8.5	534
Copper	8.9	556
Cast iron	7.2	450
Steel	7.8	487
Lead	11.3	710
Zinc	7.1	440
Cork	0.24	15
Cedar	0.35	22
Mahogany	0.7	44
Maple	0.53	33
Oak	0.77	48
Pine	0.48	30
Spruce	0.45	28
Teak	0.99	62
Styrofoam	0.02	1.3
Concrete	2.3	144
Water, fresh	1.0	62
Water, sea	1.02	64
Diesel fuel	0.88	55
Kerosene	0.8	50
Gasoline	0.72	45

Goderich 35

This Goderich 35 is a soft-chine steel yacht designed by Brewer, Wallstrom and Associates. The hull was built by Huromic Metal Industries of Goderich, Ontario, and finished and owned by René and Huguette Villeneuve of Montreal. Although it looks like a fiberglass boat, the crew knows that the strength of steel stands between them and any alien objects. With a displacement of 17,000 pounds and a sail area of 649 square feet, she has good performance to go with her attractive lines.

4

Electrolysis, Corrosion, and Painting

One problem that must be examined closely in a metal boat, which is only a minor problem in boats of other material, is *electrolysis*. Whenever two different metals are joined electrically and placed in a conducting liquid, an electrical current will flow through the conductor and one of the metals will be eaten away.

Figure 4.1 shows a simple experiment with a zinc plate and an iron plate in a container of saltwater (sodium chloride). This is called a *voltaic* or *galvanic cell* (depending on whether you are a disciple of Volta or Galvani). If an ammeter and a voltmeter are connected across these plates, a current and a voltage will be read. The voltage will depend on the material in the plates and the current will depend on the resistance in the circuit. If we allow the system to run for a while, the zinc will slowly disappear.

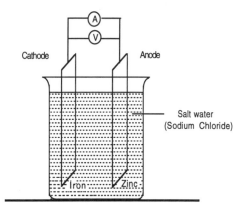

Figure 4.1 An electrolytic cell.

This is because the zinc plate gives up electrons that flow through the connecting circuit to the iron plate. When the zinc gives up its electrons (two per atom), it becomes positively charged and combines with the chloride in the salt-water. The iron plate becomes negatively charged, due to the flow of electrons through the circuit, and attracts the positively charged sodium from the solution. If dilute hydrochloric acid (HCl) is used instead of the salt solution, and lead is used rather than iron, we have a lead/acid battery like that used in automobiles and boats. The action at the zinc anode will be the same, while at the lead cathode the hydrogen from the acid (H^{++}) will pick up electrons from the cathode and become gaseous hydrogen. This will form tiny bubbles on the surface of the cathode, and when they become big enough, they will rise to the surface and bubble off as free hydrogen. This is the danger with batteries, so good ventilation is required around the top of a battery box to allow this explosive and lighter-than-air gas to escape.

What happens after a good day's sailing, when you return to the marina in your aluminum boat, plug in the shore power, and sit back and relax? If your shore electrical system is grounded to the hull, the ground wire will form an

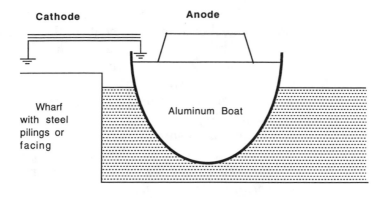

Figure 4.2 Shore power grounded to hull.

electrical connection with the shore and you will be sitting in the anode of a large galvanic cell (Figure 4.2). While you wait for the sugar to dissolve in your coffee, your boat will be slowly dissolving into the sea. This is a critical problem in aluminum boats, due to the large amount of steel used in wharves and ships. Any electrical contact between an aluminum hull and shore can prove disastrous if maintained for any length of time. The same problem, though less acute, exists for a steel hull if it is grounded to the shore. The solution to this problem is to make sure that the shore power is not grounded to your hull. This can be done by bringing the shore power into the boat via a 1:1 transformer, and grounding the boat side of the transformer to the boat, but not the shore side. When you've done that, there is likely to be a small potential difference between the shore side of your shore plug and the boat, so beware of electrical shocks.

A still more disastrous condition can occur if there is a short circuit in your shore power system. If one or two volts of galvanic voltage can slowly eat away at a hull, you can imagine the damage that could be caused by an impressed voltage of 110 volts.

The problem of electrolysis must be constantly considered when building a steel or aluminum boat. To the extent possible, a single metal should be used for everything. If you are building an aluminum hull, make sure that the cleats, stanchions, windlass, winches, water tanks, etc., are all made of aluminum, and you will avoid many problems. If it is necessary to join two metals, say, aluminum winches on a steel deck, ensure that there is an insulating pad of nonconducting material between the two. A stainless steel propeller should be considered rather than the more common bronze one, especially on an aluminum boat.

The electrochemical table (Table 4.1) shows a list of metals in order of their activity, with the more anodic or posi-

Table 4.1　Electrochemical Series of Metals

The most active metals are at the top of the table and these will "sacrifice" themselves to the metals lower down the table.

Zinc
Aluminum
Cadmium
Mild steel
Stainless steel, active
Brass
Aluminum bronze
Tin
Copper
Bronze
Copper-nickel alloys (90/10, 80/20)
Lead
Copper-nickel alloy (70/30)
Nickel
Stainless steel, types 302 and 304
Stainless steel, type 316
Titanium
Graphite

---

tive at the top, and the more cathodic or negative at the bottom. The important information from this table is that the higher up in the table a metal is listed, the more active it is. If it is connected to a metal that appears lower in the table, the higher metal will be sacrificed to the lower. That is why zinc is commonly used as a sacrificial anode: it is more active than aluminum or steel and will go into solution before the main hull material. The second item of information from the table is that the further apart two metals appear in the table, the stronger the electrical driving force and the faster the higher metal will disappear. Carbon (graphite)/zinc batteries are common dry-cell batteries because the two materials have the largest potential difference or voltage between them of commonly available materials. Nickel/cadmium (NiCad) batteries supply a lower voltage because the two materials are closer together in the electrochemical series. This, incidentally, is why NiCad batteries cannot be used on some voltage-sensitive equipment.

Even a single piece of metal can have tiny galvanic cells formed on it due to different types of metal in its alloy. Brass, for example, is made up of copper and zinc, and in seawater, a galvanic cell is set up in which the more active zinc is dissolved, leaving a weak porous part. This process is called *dezincification*.

Electrolytic cells can be set up even in uniform metal by a variation in oxygen concentration: under a drop of water, the metal at the center of the drop will have access to less oxygen than the metal at the edges and will become anodic relative to the outer edges of the water drop, producing a small rust spot at the center. These oxygen concentration cells can also be found in *pit corrosion:* any pits left on a surface, say, at the edge of a poor weld, would cause oxygen concentration corrosion when moisture entered the center of the pit. This would in turn deepen the pit and exacerbate the problem. A piece of wood on a steel plate would have a similar effect: moisture would be trapped under the wood and access to oxygen cut off, furnishing all the requirements for a galvanic cell, that is, potential difference and an electrolyte.

Stainless steels are subject to oxygen-deficiency corrosion. If these steels are open to the air, an oxide layer will form on them and prevent further corrosion. But if they are denied air, for instance, under a cover that keeps out air but lets in moisture, the steel will react very much like mild steel. That is why stainless steel is not suitable for underwater valves and is shown on most electrochemical tables in two different positions, active and passive.

Another cause of corrosion is excess stress in an area. Stress corrosion is caused by concentration factors such as a sharp corner in a hatch or other hull opening. The local variation of strain (distortion) will produce a potential difference even in uniform metal. To avoid this problem, ensure that all internal corners are rounded. This will make the boat stronger and stress corrosion will be avoided.

How then do we build a boat that is not going to be beset by electrolytic corrosion? First, we must ensure that we do not use different metals that are in contact with one another electrically in a moist environment (Figure 4.3). Hull valves can be made out of reinforced plastic, rather than the bronze that is more commonly used. The reinforced plastic valves have a plastic shaft, which tends to be a weak point. If the valve becomes hard to open, say, due to not being used for a while, it is easy to finish up with a valve handle and broken shaft in your hand. However, these valves are worth using on an aluminum boat. The cutlass rubber bearing on the propeller shaft should not have a brass outer casing; instead, it should be all rubber. Stainless steel propellers are available and, even though they are not as malleable as the normally used bronze propellers, they are worth considering from the point of view of corrosion prevention. Plumbing inside the boat should be plastic rather than copper, unless it is under pressure or intended for a fuel system.

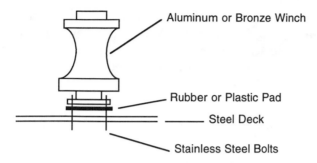

Aluminum or Bronze Winch

Rubber or Plastic Pad

Steel Deck

Stainless Steel Bolts

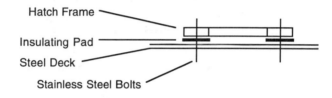

Hatch Frame

Insulating Pad

Steel Deck

Stainless Steel Bolts

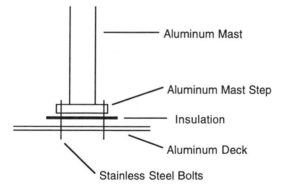

Aluminum Mast

Aluminum Mast Step

Insulation

Aluminum Deck

Stainless Steel Bolts

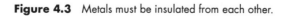

Figure 4.3 Metals must be insulated from each other.

When two dissimilar metals must be used, they should be well insulated. Put a rubber pad between an aluminum winch and a steel deck, the same with aluminum hatches or portlights on a steel boat, and under the mast.

SACRIFICIAL ANODES

Even after carefully avoiding all obvious sources of electrolytic corrosion, it is still wise to protect the hull with sacrificial anodes. These are pieces of metal attached to the hull and made of a more active or anodic material than the hull. Zinc is normally used, and preformed streamlined zinc blocks are easily available from marine stores to bolt onto the hull, or attach to the propeller drive shaft. If the boat is to be left alongside for a long period of time, the hull-borne sacrificial anodes can be saved by hanging large zinc anodes over the side on the end of a wire attached to the hull. These extra anodes can be taken onboard or left ashore when you sail.

PAINTING

Steel

A vast amount of research has been done on the preservation of steel hulls. All the navies of the world, and all the operators of merchant ships, have invested money into determining how best to reduce the cost of preventing corrosion in their ships by using various paint systems. The basis of the more effective systems is a zinc coating covering the entire hull, inside and out. After this zinc primer, other layers of paint are built up to provide mechanical protection, then an outside layer to provide good looks and protection from sunlight above the water and protection from fouling below the water.

The two types of zinc primer are organic and inorganic, which refers to the liquid part of the mixture. Inorganic primer provides better electrical conductivity and better protection against electrolytic corrosion.

After the hull has been built and all welding is finished, the hull must be sandblasted, inside and outside, down to clean shiny metal before it can be painted. This is an expensive and dirty job but there is no substitute for it. Any attempt to chip, scrape, or pickle the hull will just result in a hull that is always "weeping" rust stains. The sandblasting and surface preparation must also be done on the inside of the hull, because the problems associated with moisture and rust on the outside of a steel hull are just as bad on the inside. After the steel is taken down to bare shiny metal, it must immediately be given a coat of inorganic zinc primer, say, within half an hour, to prevent a thin layer of rust from forming on the exposed metal. Don't wait until the next day or you will have wasted a lot of time, money, and sand. Different paint manufacturers have different paint systems, so you should stay with one manufacturer for your primer, undercoat, and final surface paint to ensure each coat will bond to the other.

With the whole hull covered with zinc, inside and out, we finish up with a huge sacrificial anode enveloping the hull. If there are any surface breaks in the paint system, due to chips or scratches, the zinc will protect the steel and prevent corrosion. On top of the primer, a thick layer of protective paint is built up, and then the final surface paint.

Another way of zinc coating the boat is with a hot zinc spraying. In this process, zinc is melted and blown onto the steel surface. Not many places will have this equipment available, and I am not convinced that it is worth the extra cost and difficulty.

The system recommended by International Paint Ltd. consists of an inorganic zinc primer (zinc silicate), a high-build epoxy undercoat, and a two-part epoxy final paint. This primer is quite thin and easy to apply with a paint sprayer or brush. The next layer of high-build epoxy is thick and needs to be put on with a roller. The final coat of two-part epoxy is difficult to apply. Two-part epoxy not only has to be diluted with great precision, it is also very sensitive to temperature, and you will continually find the weather either a bit too warm or too cold. So read the directions carefully. When applied correctly, it provides a beautiful and tough surface.

Below the waterline, you must combat fouling as well as corrosion. Grass, algae, and barnacles will swim for miles just to establish their home on your nice new hull. Commonly used antifouling paints for fiberglass and wood hulls use copper as their active ingredient (the sailing ships of 200 years ago used to apply copper sheets; now we use copper paints), but copper should not be used on steel or aluminum because it will react with the hull to cause electrolytic corrosion.

Tin-based paints have commonly been used on metal hulls in the past, but they have now been banned in some countries, so you will have to check with your paint supplier for the latest miracle paint. One thing all these paints seem to have in common is high cost.

When comparing the cost of aluminum and steel boats, one should not omit the cost of sandblasting and painting the steel. These costs can be as high as 10% of the cost of the bare hull. When painting is taken into consideration, the cost difference between steel and aluminum rapidly diminishes.

Aluminum

Many aluminum fishing boats and workboats are simply left with their upper works bare. The tight skin of aluminum oxide protects the metal from corrosion, and the work and time needed to paint the boat are more advantageously spent catching fish. However, most pleasure boaters prefer to paint their boats for appearance's sake. Still, you don't have to paint the inside, which saves hundreds of dollars and many days of work.

A typical aluminum paint job consists of mechanically cleaning the surface by grinding or sandblasting, then acid washing the surface to prevent oxide formation. The U.S. Paint™ system uses Alumiprep® 33 for the acid etching, then the surface is washed and Alodine® 1201 is immedietely applied before the surface dries. When the Alodine dries, the anticorrosive primer is applied, in this case Awlgrip® 545. After this, a high-build undercoat is applied to provide protection for the primer and to fair the surface before the final topcoat.

Each paint company has its own system of paints that can live together chemically. So choose one manufacturer and stick with its recommended system. Don't mix and match.

Puffin, *by Ted Brewer*

Puffin is a 36-foot aluminum motor sailer. She has three keels, giving her a very shallow draft. In areas of high tides, she can sit on the bottom at low tide to allow for a quick bottom clean or a change of zinc anodes, and she can be hauled without needing a cradle. Her "radiused-chine" building technique provides an easy-to-build boat that has a smooth, rounded look. The pilothouse offers welcome protection in colder climates, where you can enjoy your sundowners and watch the rain beating on the large windows.

L.O.A.	36 ft, 5 in.	Displacement	25,500 lbs
L.W.L.	33 ft, 11 in.	Ballast	7,500 lbs
Beam	12 ft, 1 in.	Sail area	771 sq ft
Draft	4 ft, 0 in.		

CHAPTER

5

Welding

Your steel or aluminum boat will be built by cutting metal into various odd shapes, then welding them together to make a watertight container, which is reinforced on the inside by frames and stringers. The two main types of welding are *gas welding* and *electric welding*. The goal of this chapter is not to teach you how to weld, but simply to present some general information on welding. If you are going to weld your boat yourself, you will need to take a welding course, easily available both during the day or in the evening. Most people would do better to hire a welder with an inert gas welding outfit to do the interminable welding of thin plates that form the hull than tackle such a job themselves. That way, they will have good-quality welds and a strong and smart-looking hull.

In gas welding, a very hot flame is produced by burning acetylene in oxygen to melt the metal. The molten metal from the two pieces to be joined then flows together to form one continuous piece of metal. A rod of similar material is usually melted in this pool of liquid metal to fill in any gaps between the two pieces to be joined. In electric arc welding, a high electric current flows from the welding rod to the material being welded, melting the pieces to be joined and the welding rod into a pool of liquid metal, which then hardens into a continuous piece of metal.

GAS WELDING AND CUTTING

Oxy-acetylene welding is the oldest type of welding and can be used for practically any type of metal, from steel and stainless steel to aluminum. An experienced welder can produce good welds using this type of torch with a great variety of materials, but it is much slower and more difficult to use than an electric arc welder. The most common use for an oxy-acetylene torch in a small boatbuilding shop is to heat and bend or to cut metal.

A slightly different type of torch is required for cutting metal. In the normal welding torch, the oxygen and acetylene are controlled by two valves on the torch body. By adjusting the flow of each gas using these valves, the correct flame can be obtained for melting metal. A cutting torch has the same two valves plus an extra spring-loaded lever, which, when depressed, provides an extra flow of oxygen. Steel and aluminum will both oxidize in the presence of oxygen, and the speed of oxidation depends on the amount of oxygen present and the heat of the metal. Very rapid oxidation is, in fact, burning, and this is what happens when metal is cut using an oxy-acetylene torch.

The nozzle of the torch has several small holes around its outer annulus, or ring, and a larger hole in the center. The hot oxy-acetylene flame is produced by the gases passing through the smaller holes. This flame is applied to the metal to be cut until it is white-hot, then the cutting oxygen lever is depressed, providing a large flow of oxygen through the larger hole. When this rich supply of oxygen hits the hot metal, the metal starts to burn and will continue to burn as long as the oxygen supply is maintained. With practice, a clean cut can be made, even in metal that is quite thick. A burning-torch provides a quick, easy, and inexpensive method of cutting metal.

ARC WELDING

Electric arc welding, or stick welding, is the most common method (Figure 5.1). The machines used can vary from inexpensive ones powered by a 120- or 220-volt home circuit, to more expensive machines that can use large welding rods continuously. For the home boatbuilder, a light-duty inexpensive machine in the 150- to 200-amp range will be large enough, since the metal being welded is, for the most part, thin plate. You will spend more time cutting and fitting plates than welding them together, giving the equipment lots of time to cool between welds.

Welding machines are rated at a 60% usage rate; that is, a 100-amp machine should be able to be used at its full 100 amps for six minutes out of ten. The time off allows the ma-

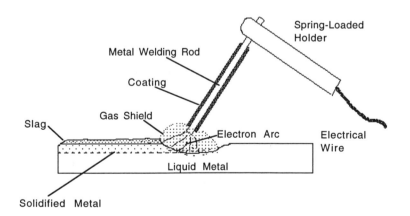

Figure 5.1 Arc welding.

chine to cool down, and the welder to change welding rods, admire his work, stretch, and plan his next bead. (This is a standard set by the American National Electrical Manufacturers Association, and it may vary in other countries.)

When reading advertisements for welding machines, read the small print carefully. I have a catalog that advertises a 230-amp arc welder and the detailed description reads:

> Welder operates at 100% duty cycle to about 100 amps, tapering off to 20% duty cycle at full amperage.

So this "230-amp" machine is really about a 150-amp machine at the 60% cycle. At least this distributor is honest about the machine's abilities; others might simply advertise it as a 230-amp machine, and the unfortunate buyer would wonder why it kept overheating with continuous heavy use.

In addition to AC welding machines, there are also DC machines. These have advantages in some situations, for example, by changing the polarity of the machine, that is, changing the work from positive pole to negative pole, the amount of heat flowing into the workpiece can be varied. If you are welding in the overhead position, with the weld on the bottom of the pieces to be joined, it would be better to have less heat in the welded part so that the weld would set faster. However, since the less expensive machines are usually AC, they are the ones that the amateur welder normally uses.

Once the metal to be welded is melted, we face the problem of how to stop it from combining with the oxygen that makes up 21% of the atmosphere. If the molten metal is left exposed to the air, it will rapidly combine with oxygen to form a weak mass of oxides. The nitrogen that forms the remainder of the atmosphere is also a danger to our pool of metal. This nitrogen is absorbed by the molten weld but, as the metal cools, it comes out of solution to create porous pockets in the metal or to form nitrites, which make the weld hard and brittle. The reduction in elasticity of the weld caused by these nitrites increases the possibility of the weld cracking. When the weld is under high pressure, it will crack rather than yield.

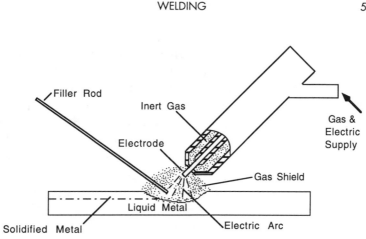

Figure 5.2 Tungsten inert gas welding.

To protect the hot weld from the air, a coating is placed on the welding rod. This coating, or flux, melts and some of it vaporizes to form a protective cloud of inert gas around the molten pool of metal. As the metal cools, the molten flux forms a slag over the weld to isolate it from the air. The slag is chipped off after the metal has cooled, leaving clean metal below it.

INERT GAS WELDING

In gas-shielded welding, an inert gas is used to blanket the molten metal during the welding process in order to prevent oxidization. No flux is used, so the final weld has no slag on it, and as soon as the electrode and welding rod are removed, the weld is clean. Also, the chance of inclusions (dirt or particles in the slag) occurring in the weld is reduced. The two main types of gas-shielded welding are tungsten inert gas (TIG) and metal inert gas (MIG).

In the TIG welding system (Figure 5.2), a tungsten electrode is held in the center of a hollow welding gun. An inert gas flows between the tungsten electrode and the outside of the gun. The electrode melts the base metal, and the pool of molten metal is enveloped by the inert gas, which prevents oxygen and nitrogen from contaminating the metal. The inert gas used most often is argon. Argon is heavier than air and does not dissipate as fast as helium, another inert gas that is sometimes used in welding, but which is lighter than air. Various gas mixtures are used for different types of metal.

To build up a weld, a metal rod of appropriate material is introduced into the pool of molten metal. Because the tungsten itself is not consumed, this is a very easy method of welding since there is no need to continuously adjust the gun's distance from the metal. In normal arc welding, where the welding rod melts, one's hand must constantly adjust for the disappearing welding rod.

The tungsten electrode can also be used for cutting metal, but leaves a poor-quality edge, because the metal is melted rather than being burned as with an oxy-acetylene torch, and this leaves molten blobs that harden along the edge. Also the gap, or *kerf*, that is cut out is wide, thus wasting material.

The MIG welding system consists of a similar gun with a flow of inert gas, but instead of a nonconsumable tungsten electrode in the center of the gun, there is a metal rod of the material suitable for the metal being welded (Figure 5.3). This rod is melted at the tip and becomes part of the weld. As the rod is consumed, it must be continuously fed through the center of the gun. This is done automatically by a machine that holds a roll of welding wire and feeds it at the required speed through the welding gun.

Inert gas welding equipment is more expensive than the more common arc welder, plus there is the extra cost of the cylinders of gas. The cylinders are usually rented from the gas supplier and the gas is bought. However, this system makes such clean welds that it should be seriously considered even by the amateur building his first boat. In the total cost of the boat the welder is a very small percentage, and good welds can make the difference between a professional looking hull and one that looks homemade.

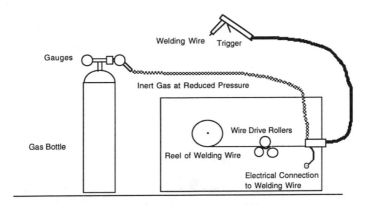

Figure 5.3 Metal inert gas welding.

Before you can start building your boat, you will need to study and practice welding for many hours. There are many good books on the market and courses are available in night schools. I would suggest that the minimum equipment you need is an oxy-acetylene cutting torch and an electric arc welding set. If you can afford it, a MIG set will make for faster and better welds and is a must for aluminum boats.

KEROSENE TEST

After you have completed a weld, how can you tell if it is watertight? One method would be to wait until you have finished the boat, then fill it full of water to see if there are any leaks. Not a very practical method since a 35-foot sailboat would hold 40 to 50 tons of water. An easier method is to put a thin (low-viscosity) liquid on one side of the weld, and check the other side to see if capillary action has drawn any liquid through the weld.

The simplest procedure is to dry off one side of the weld, say, with a propane torch, then put a few drops of kerosene on the other side. Any voids in the weld will quickly show up as a moist patch on the dry side. This test can be made more sensitive by adding a fluorescent material to the kerosene and checking the dry side with a black light.

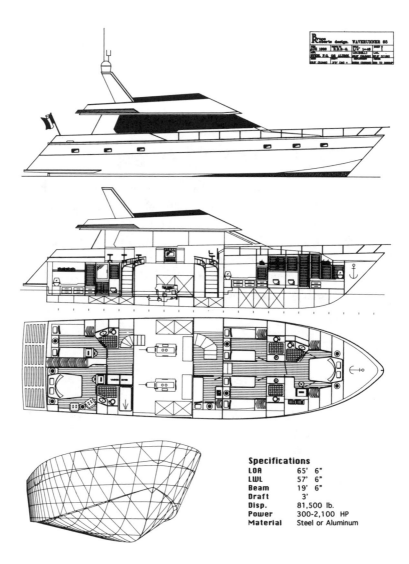

Specifications

LOA	65' 6"
LWL	57' 6"
Beam	19' 6"
Draft	3'
Disp.	81,500 lb.
Power	300-2,100 HP
Material	Steel or Aluminum

Bruce Roberts' Waverunner 65

CHAPTER

6

Lines

The lines drawing of a boat attempts to represent the complicated three-dimensional shape of a hull on two-dimensional paper. To accomplish this, certain standard formats have been developed. These "lines" are like a language, enabling the naval architect to communicate his or her ideas to the builder. Lines drawings show the shape of the boat and are accompanied by a table of measurements that allows the builder to enlarge the small drawings into full-size patterns. This reproduction of the plans to full size is called *lofting*, probably because the most easily available large open space in which to lay out the full-sized plans was often the loft of the building shed.

The naval architect is not expected to build full-sized models of designs before selling them. So for *one-off designs* or new designs, when your boat is built it might be the first time that these plans have been produced as a full-sized three-dimensional model. This means that even though you carefully copy the measurements from the table to your full-sized frames, when you get ready to draw a smooth curve through the points, the points may not lie exactly on the curve. When this occurs, you will draw a smooth curve through the majority of the points. When the full-sized frames are set up in the building jig, a long batten must be laid along the length of the boat, across the frames, and adjustments made so that it just touches each frame. This process is called *fairing* the lines.

STATIONS

The first step in building your boat will be to lay out the boat's stations full size on sheets of plywood. The frames are then built to fit these patterns. The length of the boat is imagined to be divided into sections by planes cut through the hull athwartships. These sections are called *stations*. In Figure 6.1, stations 0 to 6 are equally spaced, and the stern (after plane) and bow (fore plane) are located as required for the final shape; often the station spacing will be reduced in the stern and bow area to help describe the more rapidly changing curves in these areas of the hull.

If the boat were to be sliced into sections along these stations, the sections would look like those shown in Figure 6.2. However, because both sides of the hull must be the same, it is necessary to draw only half of these sections. This allows the naval architect to portray the boat from its widest beam forward on one side of the body plan and from the widest point aft on the other side.

WATERLINES

To be able to plot the curved lines of the half-sections, you must place a square grid over them. This is done by means of *waterlines* and *buttocks*. For the waterlines, the designer

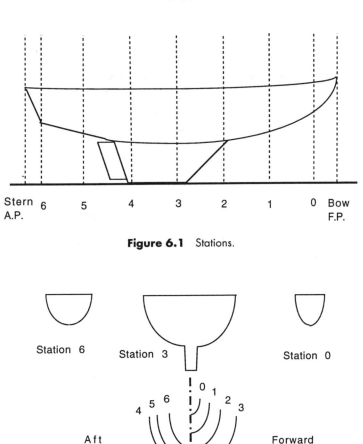

Figure 6.1 Stations.

Stern 6   5   4   3   2   1   0 Bow
A.P.   F.P.

Station 6   Station 3   Station 0

              6
          5   0 1
        4   2
                        3

Aft   Forward

Figure 6.2 Cross-sectional view of stations.

calculates the location of the waterline at which the boat will float with no stores or fuel on board. This becomes the *design waterline* (DWL). Other horizontal planes are then imagined through the hull at equal spacing above and below the DWL. These waterlines might be numbered from the baseline up or might have a different numbering system above and below the DWL (Figure 6.3).

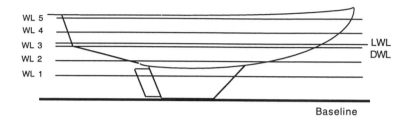

Figure 6.3 Waterlines.

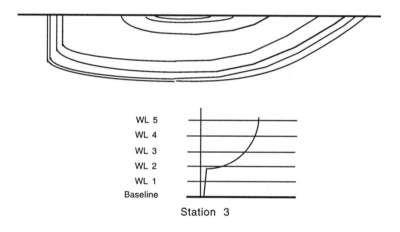

Station 3

Figure 6.4 Waterlines.

The second important waterline is the *load waterline* (LWL). This is the line at which the boat will float when normally loaded with fuel and water.

When seen in the plan view (Figure 6.4), these lines give a good idea of the shape of the boat. In the half-breadth body plan, waterlines appear as horizontal lines (Figure 6.4). By knowing the vertical spacing of the waterlines, the curve of the section can be plotted by one series of numbers that shows the distance of the hull from the centerline at each waterline. These distances are called *offsets*, because the distance that the hull is offset from the centerline is given.

BUTTOCK LINES

Buttocks are imaginary vertical planes that cut through the hull fore and aft (Figure 6.5). The planes are equally spaced from the centerline. When drawn on a side view of the boat, buttock lines help to show the shape of the hull in a similar fashion to waterlines in the plan view. In the lines drawing of the half-breadths, the buttocks appear as vertical lines. Points on the hull can be located by giving a measurement from the baseline to the intersection of the buttock and the hull (Figure 6.6).

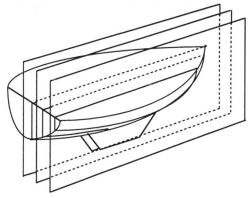

Figure 6.5 Buttock lines.

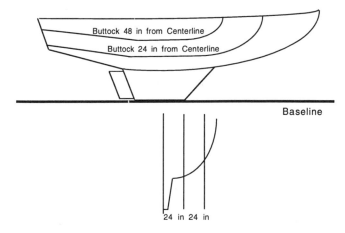

Buttock 48 in from Centerline

Buttock 24 in from Centerline

Baseline

24 in 24 in

Figure 6.6 Buttock lines.

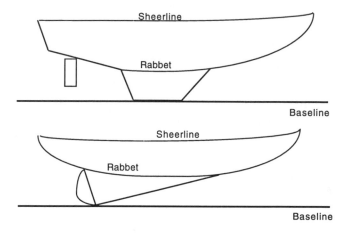

Figure 6.7 Profile.

PROFILE

Two remaining important lines are the *sheerline* and the *profile* of the hull. The sheerline is the line formed by the intersection of the deck and the boat's hull, as seen from the side of the boat (Figure 6.7). In older wooden boats there was a pronounced curve to this line, partially due to the fact that the hull was built with parallel planks. When these were brought together at the stem and stern, they naturally curved upward. In more modern boats the sheerline can be raised in the center of the boat so that it is much straighter, resulting in more headroom inside and a safer boat with more reserve buoyancy. A perfectly straight sheer is rather ugly, however, so some curve is usually kept for looks; even a reverse sheer has more eye appeal than a straight line.

The profile measurement will establish the curve of the bow and stern, and the bottom of a sloping keel (Figure 6.7). A bow with a pronounced curve in its profile will often have a separate drawing showing its curve offsets from a reference line, such as FP, with additional waterlines shown to define the curve (Figure 6.8). Most modern sailboats and powerboats have straight or almost straight stems because they are easier to make.

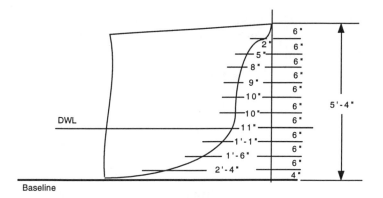

Figure 6.8 Stem curve.

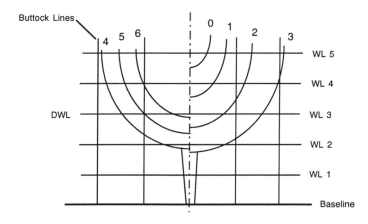

Figure 6.9 Half-breadth lines drawing.

HALF-BREADTH LINES AND TABLES OF OFFSETS

We now have sufficient information to be able to easily read the half-breadth lines drawing (Figure 6.9). This shows the shape of the forms required at each station. Because both sides of the boat are the same, only half of the shape is shown for each station. The drawings show the shape of the station forms, but the important information required to produce these shapes in their full size is shown in the *table of offsets* (Figure 6.10). This table gives the full-sized offsets from the centerline and heights above the baseline, which will describe the hull form at each station.

These lines must now be laid out full size. To do this, take several sheets of plywood and attach them so that they form a flat surface big enough to draw the lines full size. The plywood can be reused later, during the construction of the inside of the boat, for lockers, etc.—it won't be wasted. Draw a grid on the plywood that consists of the centerline, waterlines, and buttocks. Then carefully measure off the distances shown in the table of offsets. If you are using feet and inches, the measurements will show three numbers, for example, 3–4–7, which correspond to feet, inches, and eighths of an inch. If the offsets are metric, the measurements will normally be shown in millimeters.

If the naval architect has provided a set of full-sized frame drawings, they will be very welcome at this point, because they will reduce your work and diminish the chance of error. However, they should still be copied onto something solid like the sheets of plywood.

When the points have been plotted on the plywood, hammer in a small finishing nail at each point and spring a batten around them. Any long flexible piece of wood will do the job. If any of the points are off the curve, it will be evident and can be easily corrected. This is the *fairing* process mentioned earlier.

One of the waterlines may be designated as the *headstock* line; if not, draw a horizontal line across the lines, so

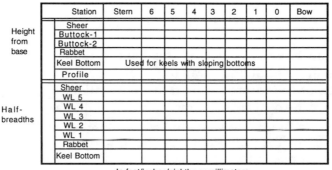

	Station	Stern	6	5	4	3	2	1	0	Bow
Height from base	Sheer									
	Buttock-1									
	Buttock-2									
	Rabbet									
	Keel Bottom	Used for keels with sloping bottoms								
	Profile									
Half-breadths	Sheer									
	WL 5									
	WL 4									
	WL 3									
	WL 2									
	WL 1									
	Rabbet									
	Keel Bottom									

In feet/inches/eighths or millimeters

Figure 6.10 Table of offsets.

that the headstock can be accurately located at each frame. This is a critical line, because it is the datum line for aligning the forms.

For a hard-chine boat, the description of the frames is much easier (Figures 6.11 and 6.12). You simply need to locate the sheer, chines, rabbet, and profile, and then join them with straight lines. The waterlines and buttocks help to illustrate the shape of the boat, but are not very useful for producing the station forms for a hard-chine boat.

These station forms may become permanent frames in the boat, in which case they should be made of the same material as the hull. In the case of frameless construction, they may simply be a means of providing the shape of the hull during construction, in which case they can be made out of lumber or metal.

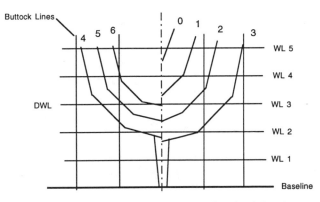

Figure 6.11 Half-breadth lines drawing for a hard-chine boat.

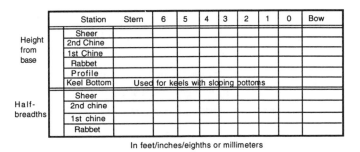

In feet/inches/eighths or millimeters

Figure 6.12 Table of offsets for a hard-chine boat.

This all-aluminum commercial fishing boat needs no painting and always looks smart. Less maintenance time means more fishing time. The light weight of aluminum allows for a full-length roof over the working area without too much topweight.

CHAPTER

7

Having the Hull Built

John had wanted his own seagoing yacht for years, and could now finally afford to have a hull built the way he wanted it. He had decided on a steel trawler-style powerboat. A small local boatbuilder gave him a reasonable estimate for the job, so John laid down his deposit and started to plan how he would finish the inside. He had bought the drawings himself and, keeping one set so he could read them at his leisure and become familiar with them, he passed the other set to the builder. The agreement was that the builder would construct the hull and decks, install the engine and fuel tank, sandblast the boat inside and out, and prime the metal inside and out.

Three months later, the builder found that the boat was costing him more to build than he had originally estimated. He had made a few mistakes, such as putting a hatch in the wrong place, and the owner had demanded that it be installed in accordance with the drawings. Finally, one day the boatbuilder called John to say "The boat is finished and I would like my final payment."

John went down the next day to look at his boat. He found it sitting in the yard with rain streaming in through the openings that had been cut for hatches and doors. Water was collecting in the bilge and the new diesel engine was exposed to the rain. Closer inspection revealed that the hull had not been sandblasted and painted on the inside as per the contract, some of the welds were leaking, and half a dozen other items were not correct. An argument with the builder ensued and, the next day, when John went down to cover the various holes in the boat with plastic to keep the rain out, he found the yard gate locked. John found a lawyer and letters were exchanged until it became evident that the builder was just not going to complete the boat.

Eighteen months and $3,000 later, John won his court case. And the builder? He declared bankruptcy and sold John's boat (in which John had invested $20,000) for $15,000 to pay off his debts. John was left with nothing. So what type of protection is available to the buyer in a case like this? The only protection he or she has is to hold back the final payment. If you hold back 100%, then you have 100% protection. If only 10% is held back, then you have practically no protection.

Therefore, an important discussion with the builder will be the schedule of payments. The builder will want you to pay as much as possible before work is started and the remainder during construction. Your goal, however, should be to keep back as much as possible until your final acceptance of the hull.

DRAWINGS

Your first step in building or buying a hull will be to choose a design and then purchase a set of drawings and specifications from a naval architect. Some naval architects design for the home builder. Their drawings will often show sev-

eral ways of building certain parts of the boat, and the specifications will usually be rather general, with many alternatives shown. This is what you need if you are building the
hull yourself, since you will want to choose the shape and
method of construction most suitable to your needs and abilities. The material and parts chosen for the construction process are often decided by what is available rather than by
what you would really like to use. Although these types of
drawings and specifications are good for those building their
own boats, they are not satisfactory for people who are having boats built. The drawings that form part of a building
contract must be very specific, with few options, and, if any
alternatives are shown, the exact choices must be spelled out
in the contract. The specifications that accompany the drawings, often a dozen or more pages, must specify the material
and parts to be used.

A good set of drawings and specifications and a clear
contract will not protect you from a dishonest builder, but
they will prevent arguments that arise as a result of different
interpretations between an honest builder and his client. Most
builders are honest and intend to remain in business, so a
good contract and drawings will help you avoid most problems.

CONTRACT

When you start discussing a contract with a builder, he will
probably produce a "standard" contract and expect to just
fill in the blanks. Of course, this will not be a standard contract at all, but one he has developed over the years to provide the maximum protection for himself. By extension, this
means the minimum protection for the client—not what you
want at all. So scoop up the drawings, specifications, and
proposed contract and consult a lawyer. It's much more useful now, before you sign, than afterward. See if you can find
a lawyer who knows something about boatbuilding because
that will make the process much simpler. Remember, it's your

money and you are in control; don't let the builder overwhelm you.

Among the elements that should be in the contract are the following:

1. What the builder is to do; for example, construct hull and deck; install rudder; install engine; install fuel tanks, water tanks, stern tube, drive shaft; provide propeller; etc.

2. What the builder is not going to do; for example, provide hatches, rigging, stove, head, etc.

3. How the work is going to be done; specify the drawings to be used and number of sheets; specifications to be used and number of sheets.

4. If there are to be any changes to the design, both sets of drawings (yours and the builder's) should be marked and initialed so that there can be no misunderstandings.

5. The exact type and manufacturer of any equipment to be installed.

6. How much is to be paid in total and when it is to be paid. One-third when the contract is signed, one-third when the hull is completed, and one-third on final acceptance by the buyer is not an unreasonable arrangement.

7. That the buyer owns the boat, to the extent that he has paid for it, and that no liens may be put against the boat without the owner's consent.

8. That the builder is responsible for the safety of the boat until it is accepted by the owner.

9. The date the builder will start construction and the latest date for completion. A penalty clause of so much per day deducted for late completion is a great incentive for the builder to finish on time.

10. Access to the boat by the buyer while it is being built. This is the only way you can ensure that it is being built to specification. It's too late once the boat is finished. Some builders will use the excuse that their insurance will not cover you when you are in the shop; therefore, customers are not allowed in. Don't fall for it, you can always sign a waiver.

11. For steel boats, specify whether the hull is to be sandblasted and painted, inside and/or outside, and the type of primer and base paint to be applied. Visit the paint manufacturer early in the game.

Make sure the builder does not try to write a contract such that he is not responsible for the quality and correctness of the construction after you accept delivery of the boat. You will not be able to tell whether it is satisfactorily built until you have put it in the water and run it for the first time.

VISIT OTHER BOATS

Choose a builder who has been in business for several years. Then a trip to see other boats built by him will be very enlightening. Ask for the names of two or three people he has built for and visit them. If the builder says that his former customers are unavailable because they are all happily sailing in the Caribbean, beware. Most people love to talk about their boats, and a half-hour spent with someone who has just had a boat built by your builder can be the most valuable half-hour in the whole process. It will quickly become evident if the builder you have in mind is reliable, and a few extra clauses for your contract may come to mind.

Snow Crystal, *a 45-foot aluminum powerboat*

Snow Crystal spends her time in inland waterways and canals. The owner has used the light weight of aluminum to its full advantage, with a large enclosed superstructure aft. Leaving the metal bare means that no painting is needed above the waterline. The increased windage of the superstructure aft will cause the boat to head into strong winds, which is a better feature than the more common falling off of the bow when there is greater windage forward.

8

Construction Details

This chapter looks at the separate parts that, when combined, make a comfortable boat that is easy to use. A chapter could be written on each item covered here, but I have tried to show just a few of the many ways to solve each problem. If you know one or two ways of approaching each item, you can either improve on them, or simply go ahead and use the same solution that has been used before. In either case you don't have to reinvent your own wheels.

VENTILATION

A boat must be well ventilated, even when it's raining or when spray is blowing over the deck. The most common means of letting air in and keeping water out is the Dorade box. There are many versions of this ingenious device, most of them having bulkheads inside to baffle the flow of water.

75

The simplest and most effective one I have been able to devise is the one shown in Figure 8.1 with the body of the box made of the same metal as the boat and simply welded to the deck. To baffle the water and let the air in, weld a pipe through the deck as shown. The diameter of the pipe will be somewhat smaller than the throat of the air scoop, say, a 3-inch pipe with a 4-inch air scoop, thus allowing for air lost through the drainage holes. These holes should be located in each corner of the box to prevent water from collecting in the corners. The top is either a piece of wood bolted to the box or possibly a continuation of the box itself. Don't forget to cover the pipe with plastic (nonrusting) mosquito screening, which can be clamped in place with a stainless steel pipe clamp.

Figure 8.2 shows an actual Dorade box and also stainless steel bolts welded to the cabin top for a teak grab rail (in the foreground). The more items you have welded to the deck rather than bolted to it, the fewer holes there will be to leak. This is one of the great advantages of a metal boat over a fiberglass boat. With a fiberglass boat, you go to a lot of trouble to create a watertight deck, then you drill hundreds of holes in it for stanchions, toerails, winches, cleats, etc.

BULWARKS AND TOERAILS

Some type of toerail or bulwark is needed along the sides of the deck to prevent small items or large bodies from being washed under the lifelines. Figure 8.3(a) shows a method of attaching a toerail. Angle metal, about 3 or 4 inches per side, is cut into short lengths to make brackets. These are then welded to the deck and the stanchions are welded to the brackets. The toerail is then bolted to the bracket [at left in Figure 8.3(a)]. For steel boats, use stainless steel brackets and stanchions; for aluminum boats, use aluminum brackets and thick-walled aluminum stanchions.

This method of adding a toerail is really a continuation of wood and plastic boat thinking, except that the bracket is

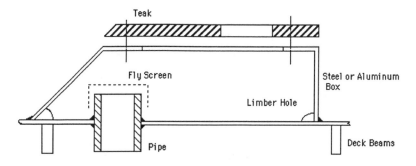

Teak

Fly Screen

Steel or Aluminum Box

Limber Hole

Pipe

Deck Beams

Figure 8.1 Dorade vent.

Figure 8.2 Dorade vent.

welded to the deck, rather than bolted, which saves four leaky holes per stanchion. A more logical method is simply to raise the sides of the boat above the deck and to cap the edge. A T-shaped piece is welded in at the deck edge, and the toerail is welded to this. An angle at the top allows a wood cap rail to be attached [at right in Figure 8.3(a)].

Figure 8.3(b) shows other methods of capping using solid stainless steel or aluminum, hollow aluminum pipes,

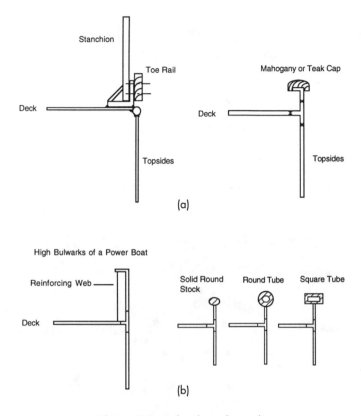

Figure 8.3 Bulwarks and toerails.

or square tubes. The top of this rail will be stepped on and ropes will be dragged over it, so it should be made of something that will not chip or rust. For high bulwarks, such as those used on powerboats, reinforcing webs will be needed every few feet to provide rigidity to the metal.

STANCHIONS

An example is shown in Figure 8.4(a) of a stainless steel bracket welded to the deck, with a stanchion welded to the bracket. A wood toerail is then bolted to the bracket. The

(a)   (b)

Figure 8.4 Stanchions.

narrow gap between the toerail and the deck (about ¼ inch) allows water to flow off the deck but stops all but the smallest items from falling overboard. The small diagonal brace acts as a useful guide for the furling line of a roller furling jib, or as a place to attach a snatch block.

In Figure 8.4(b) the continuation of the boat's side serves as the bulwark for a sailboat. The stanchion is mounted in a tube that has been welded to the deck. This allows a damaged stanchion to be easily replaced. The builder found some stainless steel door pulls in a scrap yard, welded them to a plate, then welded the plate to the deck. Handsome cleats at a few cents per pound. The hole in the bulwark would be more useful if it were taken down to the deck level where it could act as a scupper as well as a fairlead.

CLEATS

The aluminum barge in Figure 8.5 has simple and effective cleats made out of aluminum tubing. Remember, however, that cleats should be large enough so that they can take big-

Figure 8.5 Working barge.

ger lines than you normally use. It's very embarrassing when
someone passes you a ¾-inch towing line and it will not fit
on your cleat. Also, cleats should be welded to the deck, not
bolted.

RUBBING STRAKE

A rubbing strake is a useful addition to any boat to protect
the exterior finish of the hull from bumps and abrasions (Figure 8.6). A simple and effective method of construction is to
weld 2-inch angle stock to the hull a couple of feet above the
waterline. The vertical leg of the angle should point down-
ward, so that it does not become a water and dirt collector.
Wood is then bolted onto the angle metal. If this wood becomes chewed up due to a few rough alongsides, it can easily be changed. You can easily renew the paint on the wood
rubbing strip or simply leave it bare.

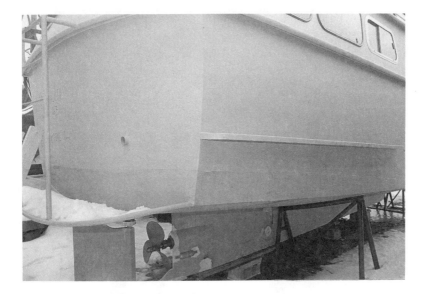

Figure 8.6 Stern view of Pedro 1000.

DAVITS

A davit is a useful item that is easy to build in metal and to weld onto a metal boat. Davits on the sailboat in Figure 8.7 enable the dinghy to be stowed at the stern and easily launched and recovered. To keep the dinghy from collecting water or scooping up large waves, it would be better to store it bottom up.

The davit on the powerboat in Figure 8.8 is used for launching and recovering a dinghy stowed on the upper deck—a good solution to an always awkward problem. If a section of the upper deck and the deck below it are made so that they can be bolted in place and removed, then machinery and heavy items from the engine room can also be easily raised and lowered, using the same davit.

Figure 8.7 Sailboat davits.

Figure 8.8 Powerboat davits.

HATCHES

Figure 8.9 shows an aluminum sliding hatch on an aluminum sailboat. The cover, or turtle, forward of the hatch prevents waves from driving water under the front of the hatch. This is a strong, weatherproof, and burglarproof hatch and its light weight makes it easy to use. Forward of the sliding hatch is a spring-loaded acrylic hatch in an aluminum frame.

The hatch in Figure 8.10 is easy to build, with a teak frame and a ⅜-inch smoked acrylic top. The beauty of a transparent hatch can best be appreciated when you have been at anchor in the pouring rain all day. The added light in the galley area makes the whole world seem brighter. On a steel boat, stainless steel tracks should be used, otherwise constant use will wear off the paint and produce rust streaks.

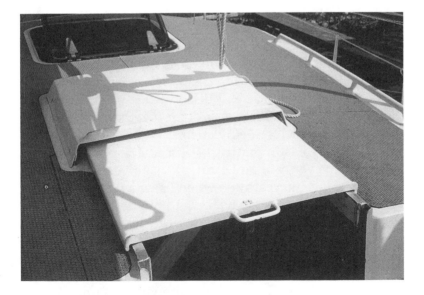

Figure 8.9 Accommodation hatch.

Figure 8.10 Accommodation hatch.

Acrylic is a strong plastic that does not become opaque with exposure to ultraviolet light. It is inert to most common chemicals, such as oil, salt spray, alkali, and most common household products. It even holds its own against acetone, as I found out when I had to clean off some spilt polyester resin. Two solvents that will dissolve this plastic, and can therefore be used to stick pieces of it together, are ethylene dichloride and trichloro-ethylene.

Another plastic that can be used for portholes and hatches is the very strong Lexan™. This polycarbonate resin first impressed me years ago when I saw a very thin extrusion in the shape of a top hat being pounded with a heavy hammer without suffering any damage. If you need it even stronger, you can get it in a glass-reinforced form. It is susceptible to ultraviolet light, though, becoming slightly opaque.

The builder of the steel boat shown in Figure 8.11 has decided to use steel hatches. Metal hatches work well with aluminum boats, but in steel they are very heavy. This cabin-top hatch will need some type of spring to help lift its weight or it will be a real finger-crusher. In a very short time, the hole in the hinge through which the stainless steel pipe passes will have its coating worn off and will start to weep rust. Solid metal hatches also make the inside of the boat as dark as a tomb.

The companionway hatch on the same boat (Figure 8.12) slides on stainless steel tracks that pass through stainless steel guides, an arrangement that is very heavy but surprisingly easy to use. This boat is really safe against vandalism because the only way to get in once it is locked is with a cutting torch.

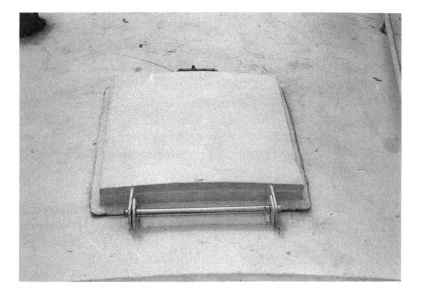

Figure 8.11 Steel hatch.

Figure 8.12 Sliding hatch.

HOLES BELOW THE WATERLINE

All below-the-water openings are potential sources of leaks
that could sink the boat, so each hole should be carefully
considered to see if it is really necessary. Underwater openings, such as suctions and drains for the heads, sinks, engine, and cockpit, must all have valves on them that are
placed as close to the hull as is practical.

The first step in planning water supplies and drains
should be to reduce to a minimum the number of underwater openings. For aft-cockpit sailboats, for example, the cockpit drains need not be a reason for two or four more holes in
the boat's bottom. The cockpit sole should be high enough
above the outside water level that the cockpit drains can simply be led aft through the transom above the waterline (Figure 8.13). This open drain system will also let propane flow
out of the bottom of the cockpit if there is a leak from your
barbecue.

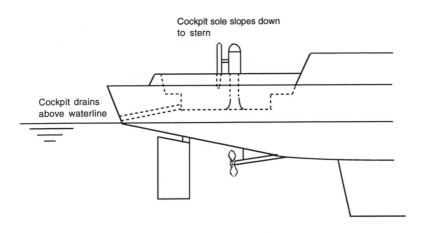

Figure 8.13 Cockpit drains.

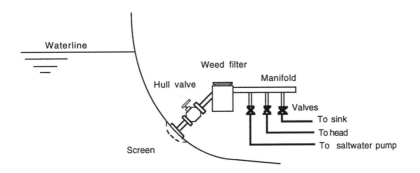

Figure 8.14 Common seawater supply.

The rudder shaft is usually another reason to have a hole in the boat's bottom. This can be avoided by using a stern-hung rudder.

Some designs go so far as to have one hull valve for all of the water supplies. This one valve leads to a manifold or seachest via a weed filter. Pipes are then led to the head, engine, etc., from this manifold (Figure 8.14). This allows the critical hull valve to be large and robust. The discharge from the engine will, of course, be well above the waterline, and the discharge from the head and sink can be just above the waterline, so there is no danger of taking in water and sinking the boat while it is alongside or at a buoy. Even these above-water openings should have hull valves on them for use in an emergency or in heavy seas.

I think the best arrangement is to have a dedicated water supply for the engine, and a single manifold supply for all of the other raw water needs.

HULL VALVES (SEACOCKS)

The area around the hull opening should be reinforced by a *doubler* plate. This is a plate, one and a half to two times the thickness of the hull, that is used to increase the strength of the hull at the hole. For a 1-inch valve outlet, for instance, a hole 3 or 4 inches in diameter is cut in the hull and the thicker plate welded in place.

Several methods of attaching the hull valve to the hull are illustrated. The first one [Figure 8.15(a)] has a plastic or bronze through-hull connector introduced into the hull from the outside. It is sealed and a large nut on the inside draws it into place. This style is commonly used on fiberglass boats. The valve is then screwed onto the through-hull connector, a tailpipe is screwed into the valve, and a flexible hose is attached to the tailpipe by means of a couple of stainless steel hose clamps.

The second example [Figure 8.15(b)] shows a valve with a flange attached. This type of valve is bolted onto the ship's side and is commonly used on larger ships where the doubler plate is thick enough to have stud holes drilled and tapped into it. The valve is then simply bolted directly onto the plate with a rubber gasket used to seal the joint.

The third arrangement, and the one that I prefer, has a pipe welded directly to the doubler plate [Figure 8.15(c)]. The inboard end of the pipe is threaded and the valve is screwed onto it. The pipe is the same material as the hull to avoid electrolytic corrosion. It is strong and can be made long

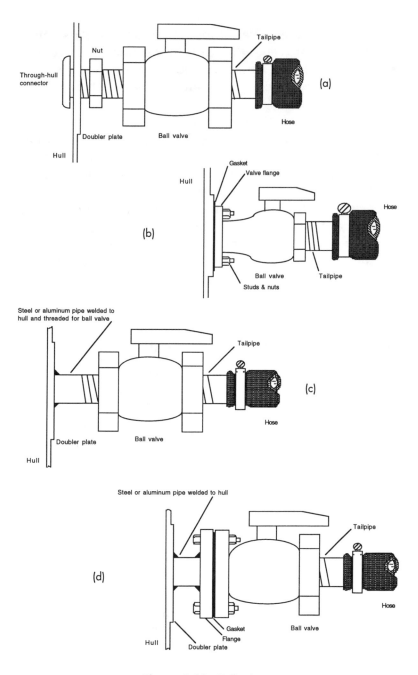

Figure 8.15 Hull valves.

enough so that the valve stands well clear of the insulation and any interior woodwork. Another similar arrangement [Figure 8.15(d)] has a flange welded to the pipe and the valve is bolted to the flange, with a watertight gasket between the valve and flange.

Ball valves are the best type for use as seacocks. These types of valves can be opened or closed by turning the handle through a quarter of a turn. They are set up so that when the handle points along the pipe the valve is open; when it points across the pipe the valve is closed. This makes it easy to see at a glance the state of the valve. Other valves, such as gate valves, can be easily jammed by debris entering the valve, they take many turns to open and close, and one cannot tell whether they are open or closed except by trying them. Even then it may not be too obvious if the valve is very stiff. Bronze is my preferred material for hull valves, although reinforced plastic valves should be considered in aluminum boats to save you the problem of electrolytic corrosion. Stainless steel valves are available but underwater they will corrode between areas with a good supply of oxygen and the oxygen-starved areas. Never use brass valves—if your supplier is not sure whether his valves are brass or bronze, find another supplier.

PORTHOLES

Although commercially manufactured opening portlights are much more expensive than simply bolting sheets of plastic onto the cabin sides, they are well worth the money. A boat can become like an oven in the tropics, and the cross flow of air from open portlights can make all the difference between simply being too warm and sweltering. Mosquito screens should be part of the assembly so that the boat can be open to the air, but closed to those insomnia-inducing insects.

Figure 8.16(a) shows a common arrangement in which the portlight frame is bolted onto the cabin side with a wood backing sandwiched between the frame and the hull. The

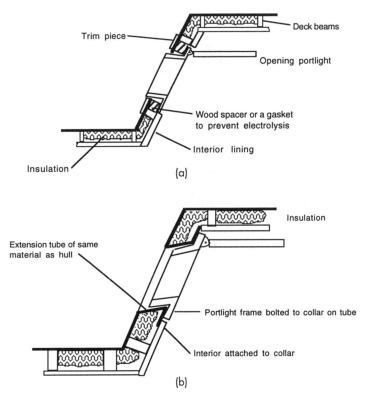

Figure 8.16 Portlights.

wood backing should be extended well past the frame to allow wood furring strips for the inside liner to be attached to it. A disadvantage of this arrangement is that rain will blow in through the opening, and there is the problem of corrosion between the outer trim piece and the hull.

A better arrangement has the frame set a few inches inboard on a metal trunking [Figure 8.16(b)]. This is more complicated to build, but allows the glass to be left open even during rainfall without water blowing into the cabin. There is no problem with leakage or corrosion around the bolts holding the portlight frame to the hull. The inner flange of the trunking should be wide enough to allow the interior woodwork to be attached to it.

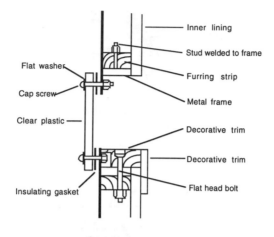

Figure 8.17 Fixed portlight.

A method of building a fixed porthole is shown in Figure 8.17. The clear plastic is simply bolted onto the outside of the cabin. A more sophisticated arrangement would have a frame around the plastic, which would spread the force of the bolts over a larger area. If a frame is not used, a large flat washer should be placed under the head of the cap screw. The inside of the opening has a metal frame welded to it, so that wood can be bolted to it. The inside lining is then attached. The top half of Figure 8.17 shows the simplest method of finishing the inside, and the lower half shows a better method, which covers the metal frame and nuts with trim that frames the opening.

Fixed portholes should be used over the chart table and navigation station. Somehow the cooling drip of seawater never seems to improve the operation of electronic equipment.

GROUND TACKLE

Nothing ensures a good night's sleep better than a heavy anchor firmly dug in, with ample scope of chain let out. For the bow anchor, I prefer a plow or Bruce anchor and chain rode. With a good windlass, it is easy to let out and recover the chain, and there is no fear of the rode chafing on the bottom or in the fairleads (chocks). Nylon rode is good for the kedge anchor, which must be light enough to carry out in the dinghy, so that you can haul yourself off if you go aground. On our nine-ton sailboat, we carry a 45-pound plow on 200 feet of ⅜-inch chain as a bow anchor, a 35-pound plow on 20 feet of chain and 200 feet of ⅝-inch nylon as a spare, and a kedge anchor on 20 feet of chain and 200 feet of ⅝-inch nylon. The kedge anchor is permanently rigged in case we start dragging through a muddy bottom.

Figure 8.18 shows a plow anchor neatly stowed on the bow roller of a sailboat. With this roller arrangement, the windlass can heave in the anchor shank until the fluke is right up against the roller. There is no need to manhandle the anchor, so anchoring and weighing anchor become simple procedures that do not require strength or agility.

Figure 8.18 Plow anchor.

Figure 8.19 shows a hydraulic-powered windlass with a Bruce anchor. This small aluminum fishing boat is typical of the north east Pacific vessels. With 300 to 500 feet of wire rope plus chain on the windlass, they can anchor in the deep waters of the British Columbian fjords.

The Admiralty pattern anchor (not shown) has been used in sailing ships for hundreds of years. These anchors look very nautical on old boats, but you need a couple of spare men on the foredeck to handle them, and if you are anchored in tidal waters, the anchor rode has a bad habit of wrapping itself around the fluke that projects upward.

One day in Atlantic City, while waiting for some bad weather to blow over, we rafted up with a friend who had a 60-pound Admiralty pattern anchor firmly dug in, and went ashore for a shower. We decided to stay ashore for dinner, and when we returned, we found that the boats had dragged halfway across the harbor. Luckily, some good sailor had gone on board and dropped our plow for us and stopped the boats.

Figure 8.19 Bruce anchor with power windlass.

Here's an aluminum sailboat with beautiful lines. These smooth three-dimensional curves require careful development by the designer, then a skilled and knowledgeable builder to produce the well-cut frames and carefully rolled sheet metalwork. Very few yards can produce a boat of this caliber.

Hull Construction

After you have chosen your design and found a place to work, the next decision will be whether to build the boat upside down and then turn it over, or to build it right side up. If you build it right side up, it will save the rather difficult job of turning over the half-finished boat to put on the deck and cabin. However, you will need a fairly substantial structure from which to suspend the frames during construction, and you will have to be constantly alert to the need to increase support for the hull and other parts of the structure as their weight increases.

If you are building outside with no roof, the upside-down procedure has the advantage of allowing rain to run off the hull as the plating is put on, whereas with the hull right side up, water will collect in many little pockets between the plating and the stringers and in a large pool at the bottom of the keel.

The boat used in this chapter to illustrate one method of construction was built upside down, then turned over to complete the deck and cabin. It is a Bruce Roberts 36, a hard-chine sailboat built of steel. It was built using frameless construction; that is, the boat was built on frames, which were removed after the hull plating was completed. This produces a boat that is strong but not cluttered with space-consuming frames.

A hard-chine boat is the simplest for an amateur builder to attempt because you don't need machinery to roll metal sheets; the only bending of metal required will be the stem and the deck beams.

TOOLS

The three procedures that are constantly repeated in building a boat are measuring, cutting, and attaching. The cutting of steel can be done in several ways, depending on its thickness and the type of edge you need. The simplest and easiest way is to use an oxy-acetylene cutting torch. This will cut through thin or thick metal without strenuous effort on your part. At first you may find your edges rather rough, but with practice you will improve.

For a cleaner edge, a cutting wheel of some type is needed. In Figure 9.1, a carborundum wheel is being used on a portable saw to cut a stainless steel chainplate—a slow job, but one that leaves a clean edge. Note the goggles and breathing mask. For aluminum, a fine-toothed carbide-tipped saw blade will do the job. Cutsaws and heavy-duty jigsaws are also useful for trimming and fitting parts.

Figure 9.2 shows a very useful tool called a *nibbler*. It punches out small pieces of metal and can be used to trim a rough-cut edge or to accurately fit a sheet of plating to the frames and chines.

LAYUP JIG

If the frameless method of construction is being used, the layup jig and frames can be made of lumber. If the frames

Figure 9.1   A carborundum wheel on a portable saw.

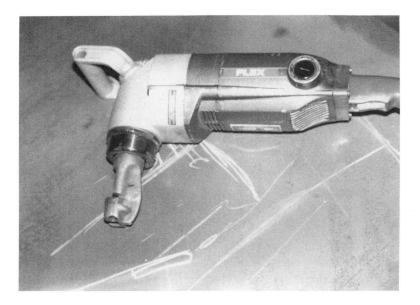

Figure 9.2   A nibbler.

are going to remain part of the boat, they will have to be of the same material as the boat. Lumber jigs must be strongly built, with 2-inch × 12-inch lumber for the main beams and strong cross-bracing to prevent twisting. The vertical supports are bolted to the main beams and have diagonal bracing to prevent movement. Figure 9.3 shows a steel jig with the vertical supports welded to the main frame.

The vertical supports, whether welded or bolted, must be accurately located at each station. As you build your jig, remember that it will have to support the whole weight of the hull (Figure 9.4). Ensure that the main beams are horizontal and are not likely to sink into soft earth as the weight increases. Ensure too that the verticals are strong enough to bear the weight of all the steel, plus a couple of builders climbing over the hull. If the vertical supports are bolted to the headstock, it will make it easier to remove the jig after the boat is turned over. One great advantage of the frameless method of construction is that the frames can be reused after they are removed. If you can find someone else who likes the same hull as you do, he or she can use the same jig and frames.

FRAMING

The frames are now built on the full-sized lines plan that you had previously laid out on sheets of plywood. One side is cut to fit the lines, then the frame is turned over and the other side cut to fit the same lines. This ensures that both sides of the boat will be exactly the same (Figure 9.5).

The frames are attached to the jig by means of the headstock (Figure 9.6). As the frame is built, the headstock is carefully located as per the drawings, with all headstocks on the same waterline. The center of the headstock is marked from the centerline of the half-breadths. After all the frames are completed, they are assembled on the jig. A wire is stretched tightly along the centerline of the jig at the height of the head-

Figure 9.3 Steel layup jig.

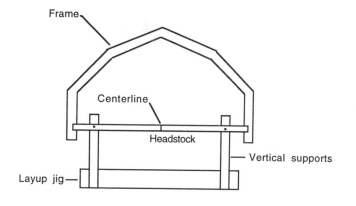

Figure 9.4 Layup jig.

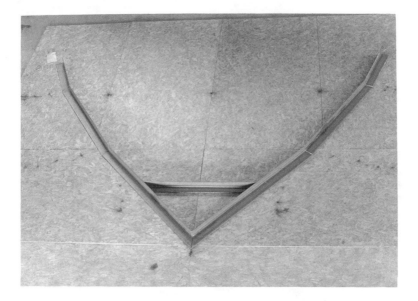

Figure 9.5 Frames.

Figure 9.6 Frame assembled on layup jig.

stocks. The centers of the headstocks are aligned using this wire, so that the center of each headstock is at exactly the same lateral position and height.

If you have access to surveying equipment, a theodolite can be useful here. Remember that, in addition to the headstocks being the same height and on the same centerline, they must also be exactly horizontal, and the frames must be perfectly vertical when viewed from the boat's side.

In our sample boat, the frames are used to give shape to the boat during construction and are then removed, the strength of the boat being in the skin, stringers, and deck frames. The stringers here are 1½-inch angle iron. This is stronger than flat stock, and it is easier to attach the interior woodwork to the inner leg of the angle iron than to flat stock. The angle iron is welded together, end to end, to make stringers the full length of the boat, then bent to fit the frames.

In Figure 9.7, the stringers have been attached to the frames using aluminum pop rivets. These are easily chiseled off when the boat is turned over, so that the frames can be removed.

After the stringers have been attached to the frames, the round stock used for the chine is carefully located. In Figure 9.7, straight edges (pieces of angle iron) are clamped to the stringers, and the chine round bar is located at the intersection of these two straight edges. It is then welded in place (Figure 9.8) using a small piece of steel that will later be cut off when the frames are removed. In Figure 9.8, note the smooth curve of the stringers and chine bar: if the stringers do not touch all of the frames when a smooth curve is formed, they should be shimmed. Make sure that the stringers are straight and that there is an even space between them. The work you do now defines the final shape of the hull, so time taken at this stage will repay you well in the final quality of the boat.

At the centerline of the boat the ends of the stringers and chine bars will be joined at the stem, profile bar, and floors.

Figure 9.7   Locating chine bar.

Figure 9.8 Chine bar and stringers.

PLATING

Plating the hull consists of making a pattern that fits the part of the skeleton to be covered, cutting the metal to fit the pattern, then forcing the metal into the curve of the boat and tack-welding it in place. The pattern can be made from thin plywood or corrugated cardboard. If you use corrugated cardboard with one side smooth and one side corrugated, you will find that it bends easily in one direction but not in the other, and that it simulates the action of a thin sheet of metal quite well.

To ensure that the skeleton is not warped, weld a plate on one side, then the equivalent plate on the other side. This will equalize the forces due to the weight of the plating and the contraction of the welds on both sides of the boat, and you will finish with a straight boat. The shell plating is tacked in place to the chine bars first (Figure 9.9). The butt weld between plates is left until last, to allow for longitudinal movement. Leave a small gap between the plates to allow for expansion and contraction during welding.

The plates are welded to the stringers, using intermittent welds. Because the welds will show through the skin slightly, even after the boat is finished, they should be evenly spaced. The evenly spaced welds can be seen in Figure 9.10. After sandblasting and painting, these welds will be almost, but not quite, invisible.

The transom will be fitted after the hull is plated. Stringers and plating will be trimmed to length, and the transom shaped to fit.

Looking like a submarine (Figure 9.11), the completed hull is ready for a cradle to be fitted. It will then take a couple of cranes to turn it over.

After the hull is turned over, the building jig and frames are removed. This is when it becomes obvious why aluminum pop rivets were used to attach the stringers to the frames. As the frames are removed, the hull becomes quite flexible, so leave several frames in the center of the boat until the deck frames have been installed in the ends of the boat. In

Figure 9.9 Plating.

Figure 9.10 Stern before transom is added.

Figure 9.11 The halfway point.

Figure 9.12, the stern has been propped up to limit unwanted movement. The transom still extends above the deck and will be trimmed to the exact curve of the deck after the aft deck frame is fitted.

Details of the bow area show the stringers and chine bars welded to the stem with triangular gussets reinforcing the bow (Figure 9.13). The deck beams are attached to vertical members that tie the deck edge to the first stringer. The neat, evenly spaced welds can be seen in this photograph. Don't worry about making this area too strong; when you hit that deadhead in the middle of the night, your first thought will not be "Did I overbuild the bow?"

The side deck beams and carlin are held in place by temporary struts that will be cut out later (Figure 9.14). Figure 9.15 shows the side deck with its long curving carlin, deck beams, intercostals, and reinforced area. The hull plating thickness is ⅛ inch, but in the way of the chainplates, stays, and anchor windlass, it has been increased to ¼ inch. This

Figure 9.12 Ready for the deck frames.

Figure 9.13 Detail of bow area. Note the evenly spaced welds.

Figure 9.14   Temporary struts hold the side deck beams and carlin in place.

Figure 9.15   The side deck, showing the long curve of the carlin, deck beams, intercostals, and reinforced area.

means that the deck beams have to be cut back by ⅛ inch to ensure that the top surfaces of the plates are level.

The cockpit area starts to take shape with the plating of the side decks (Figure 9.16). The cockpit can be prefabricated elsewhere, then fitted into place. The round bar stock at the edges of the seats makes for easier welding and an attractive rounded edge on the final seats (Figure 9.17). Make sure the tops of the seats slope in toward the center of the cockpit (or else you will end up sitting in pockets of water) and the cockpit sole slopes down to the cockpit drains. The coaming is being built around the edge of the opening.

In Figure 9.18, the cockpit—your future patio—is taking shape. Because you will spend a lot of time sitting here, both at sea and in harbor, you should make sure that it is comfortable. This means that the after-bulkhead of the cabin and the coamings should have a comfortable slope. The coamings should also be high enough to provide you with an adequate backrest and protection from the sea.

When building the coamings, don't create any closed-in boxes that are impossible to get at to paint, but where mois-

Figure 9.16 The cockpit starts to take shape.

Figure 9.17   Cockpit framing.

Figure 9.18   Detail of cockpit.

Figure 9.19 Deckhead of the cabin trunk.

ture can get in to rust from the inside. Any closed-in spaces must be airtight so that the oxygen in the space will become used up and rusting will stop.

You can see that the main hatch has been cut and framed (Figure 9.18). Make sure that this opening is big enough to make it easy to step into the cabin from the cockpit. You will be constantly going in and out of this opening and you'll soon become irritated if you have to twist or duck your head all the time. Also it should be big enough to enable you to lift out large items, such as the engine, or the stove. It's much easier to make the hole a bit bigger now, than after the boat is finished!

The deckhead of the cabin trunk is shown in Figure 9.19. If a pronounced camber is used, there will be more head-room in the center and the surface will be more rigid. This will reduce the chance of the gaunt, skeletal look of skin sagging between bones that is seen in some metal boats.

To bend the beams, weld up a jig as shown in Figure 9.20. Then, using a hydraulic jack, work along the length of

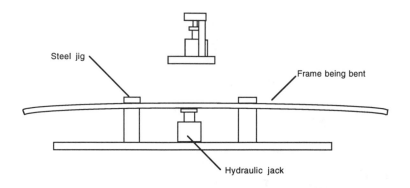

Figure 9.20 Bending the deck frames and stem.

the beam bending it a small amount each time until it matches the required curve that you have previously marked on a length of wood.

Make sure there is enough space between the cabin frames for the portholes, and that the holes can be cut where you want them—plan your bulkheads first, then the portholes. In Figure 9.21, the distance between the deck beams has been varied slightly to make space for the portholes.

In Figure 9.22, the floors can be seen securely welded to two stringers on either side and to the hull plating. Large limber holes have been left next to the keel to allow for water drainage. Limber holes should also be cut in the lower stringers to ensure that any water or condensation runs down to the bilge, rather than remaining in pockets above the stringers. The next step is to weld flat stock onto the top of the floors, so that the cabin sole can be bolted onto it.

The propeller shaft tube is fitted for the individual engine chosen. The location and angle of the tube depend on the engine being used and its location. A full-size side view of the engine can be useful in lining up the engine drive coupling and the stern tube before the tube is welded in place. In wood boats, this shaft is usually offset so that keel does not have to be cut, but in a metal boat, it can be on the centerline.

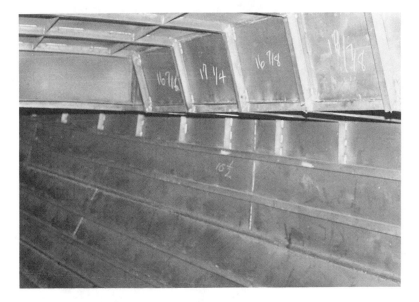

Figure 9.21 Cabin trunk frame spacing.

Figure 9.22 Floors.

Figure 9.23   Stern tube.

In Figure 9.23, the floors are being cut away to allow the engine to sit as low as possible. The floors are welded to the stringers, hull, and keel to provide a strong support for the engine's weight and the propeller's thrust.

Another practical method of lining up the propeller shaft tube is to temporarily mount the engine in its final location. Add the driveshaft and slide the stern tube and bearing onto the shaft, mount the propeller, and support the whole lot so that the weight is not hanging on the engine. You can now make sure that there is adequate clearance between the propeller and the hull and that the stern tube lines up correctly with the driveshaft. If you have a fixed skeg abaft the propeller, make sure that the shaft is angled slightly so that the shaft can be removed past the skeg. When everything is aligned to your satisfaction, weld the stern tube in place and the job is done.

In a wood or fiberglass boat, the keel is usually a large lead casting bolted to the bottom of the boat. As such, it must all be poured at the same time to prevent layering of the lead. You need a solid lump that will stay together by its own

Figure 9.24 A summer evening at the foundry.

strength. In a metal boat, with a hollow keel, the procedure is much simpler because the ballast can be put into the keel in small pieces and then some other material, such as lead or cement, can be poured around the ingots to keep them together.

If scrap lead is used, the equipment shown in Figure 9.24 is all that is needed to melt the lead and produce ingots: a camp stove, a cast-iron pot, a cast-iron cake pan, and a propane torch. In the photo, the author is spending one of many pleasant summer evenings melting the 6,000 pounds of scrap lead in the background into the neat pile of 50-pound ingots seen on the right.

These ingots were then placed in the keel and more lead was melted and poured around them to form a solid mass. If scrap steel is used for ballast, cement can be poured around it to keep it in place. The best way to finish these keel compartments is to weld an airtight cover over the ballast. This will prevent corrosion inside the keel and provide a tidy storage space.

*An aluminum tug built at the Royal Military
College of Canada many years ago and used as
a floating laboratory by the engineering students.
It is called, appropriately enough, Cordite.*

These Kort nozzles are commonly used on tugboats and are said to increase the bollard pull by 30% to 40%. This is probably due to the nozzles acting as end plates for the propellers and reducing the power lost in tip vortices. In the illustration, normal propellers are being used, but it is more common to use propellers with square ends that fit snugly into the shape of the nozzles. Another variety of this arrangement is the *azimuthing* units. In these types of units, the propeller is driven by a vertical shaft (similar to an inboard/outboard unit) and the nozzle and propeller rotate about this axis. This gives the tug great maneuverability.

CHAPTER

10

Steering Gear

The first requirement of any steering system is to give the helmsman control of the boat without requiring excessive strength; the second requirement is complete reliability.

A boat's steering gear can vary from a simple tiller directly attached to the rudder to complex wire-and-pulley systems. This can be further complicated by an automatic wind-vane steering system or by secondary steering positions.

The outboard rudder has the advantage of not having to pass through the bottom of the boat, thereby removing one potential source of leaks. Locating the rudder as far aft as possible gives the maximum leverage of the rudder on the boat, which reduces the amount of rudder angle needed and hence reduces drag. A skeg in front of the rudder gives support and protection to the rudder, increases its hydrodynamic effect and, when the boat is running straight, increases its directional stability in much the same manner as would a longer keel.

Other advantages of an outboard rudder are the ease with which a wind-vane steering system can be attached to it, and the ease with which it can be removed from the boat for repairs, even while the boat is still waterborne. A disadvantage of this arrangement is that the rudder is easily damaged if a boat runs into you, or if you back into the wall while mooring with the stern to the jetty.

The more common inboard rudder can be attached to a skeg, attached to the aft end of a long keel, or can be a spade rudder, i.e., with no support at the bottom.

The rudder attached to the end of a long keel is the strongest and best protected arrangement, but as it is further forward, it has the disadvantage of requiring more force from the rudder to turn the boat. Because of the longer keel, there is increased surface friction to slow the boat in light winds, less lift from the keel, and a large turning circle.

If we compare tiller-controlled rudders with remote wheel-controlled rudders, we see that the tiller has the advantage of quick response, less "free play" in the system than is common with wire-driven systems, simplicity, and reliability. Balanced against these factors is the possibility of locating a steering wheel farther from the rudder in a more sheltered spot. While a tiller and a wheel may require the same amount of force to turn a rudder, with a tiller you have to keep exerting this force to hold the rudder over, which can get tiring after a few hours; whereas a wheel can be held over with very little force. A good long tiller is a big help in handling a boat in strong winds, and a long tiller will also let you get further forward to huddle behind the dodger.

With wheel steering, the head of the rudder shaft should be accessible so that if there is a failure of any part of the steering gear, a tiller can be used for emergency steering: wheel-steered boats should always carry an emergency steering tiller. The head of the rudder shaft should be accessible from the upper deck level. Some boats have the rudder shaft hidden under the cockpit sole, and to get to it you have to be small and flexible and able to crawl into the cramped space. But if the steering fails, it will normally be in bad weather and crawling into a small space will cause seasickness in very short order and also leave the boat out of control. So make sure that you can get at the rudder shaft from the cockpit.

RUDDER

An example of a simple flat-plate rudder used on a long-keel sailing boat is shown in Figure 10.1. This is the cheapest and easiest type of rudder to build, and consists of a plate of metal welded to the rudder shaft, with reinforcing ribs welded to the side for rigidity and strength. It has holes drilled in the top of the trailing edge to allow for lines to be attached. In case of damage to the steering gear, the rudder can then be pulled to either side by the lines to provide for emergency steering.

The rudder has a flange at the top of the rudder shaft to allow the rudder to be removed for repairs without needing

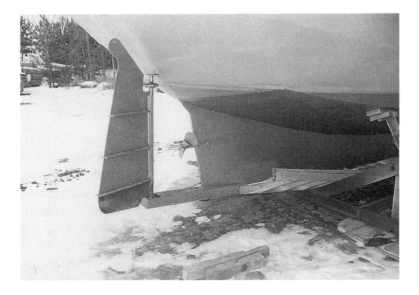

Figure 10.1 Flat plate rudder.

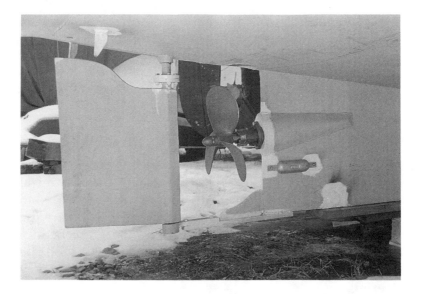

Figure 10.2 Streamlined rudder using two plates.

to drop the whole shaft out of the boat. Without this flange, the boat would have to be lifted clear of the ground, or else a pit would have to be dug behind it, to allow the rudder shaft to be dropped out of the hull. An improvement to this rudder would be to have its lower edge sloping up, so that if the keel takes the bottom, the rudder will not be damaged—an important consideration.

The rudder of the 10-meter steel powerboat shown in Figure 10.2 is more sophisticated, with two plates welded onto the rudder stock, which results in a streamlined shape. The protuberances from the hull above the rudder act as stops to prevent too much rudder movement. These are useful when going astern, since a rudder that is pushed over too far by the water simply impedes the astern movement but does not swing the stern. There is a flange at the top of the rudder shaft, and the lower bearing housing is bolted to the bottom of the keel. To remove the rudder, the lower bearing housing is removed, then the heavy rudder can be lowered

Figure 10.3 Spade rudder.

after the flange is unbolted. The large three-bladed propeller ensures that all of the engine power is transmitted to the water. Note the large sacrificial zinc anode ahead of the propeller.

Figure 10.3 shows a spade rudder on a fiberglass sailing boat. The spade rudder has the obvious disadvantage of lack of protection from debris that might be encountered in the water. It also exerts a large bending moment at the top of the rudder shaft due to the sideways force of the water acting on the rudder. This means that the top bearings must be as far apart as possible, and the rudder shaft as large and strong as possible, and even so, a blow from a deadhead will easily damage this type of rudder. Such an arrangement is satisfactory if you are just sailing out of your local yacht club or marina for an afternoon's race, but is rather fragile for exploring the far corners of the world.

The small, folding "sailing" propeller (Figure 10.3) reduces drag, but also is less efficient, reducing the engine's stopping power when you are heading for the jetty. It is more complex than a normal propeller and so is more easily damaged, and it is fairly expensive. But the reduced drag makes it worthwhile if you are racing, and is worth considering for the cruising sailor who plans to cross large oceans.

STEERING

Wire Steering

The three main types of wheel steering systems are wire, mechanical, and hydraulic. The wire system is the most common. In this system, a sprocket-and-chain arrangement, similar to a bicycle chainwheel and chain, is mounted on a pedestal. The steering wheel is on the same shaft as the sprocket gear; when the wheel is turned, the sprocket pulls the chain in one direction or the other. The chain is connected at its ends to wires that run down the pedestal, pass over pulleys and head aft (or forward), to the rudder.

Two common arrangements that enable the pull of the wire to rotate the rudder shaft are shown in Figures 10.4 and 10.5. The *quadrant* arrangement of Figure 10.4 has the wires passing over two pulleys close to the quadrant to change their direction by 90 degrees. They then pass over two grooves on the quadrant, which is bolted to the rudder shaft. The steering wheel can be located ahead or astern of the rudder shaft. The pulleys under the base of the steering wheel pedestal can be on a different plane than the quadrant, and the quadrant pulleys will allow for the misalignment. If the path to the steering wheel is longer and more complicated, as in a center cockpit boat, you will need more pulleys to guide the wires. But the more pulleys you have, the more resistance there will be in the system, and the more loose play in the wires. So keep the run as short and simple as possible.

In the second arrangement, a *disk drive* arrangement, you do not need the extra pair of pulleys (Figure 10.5). The

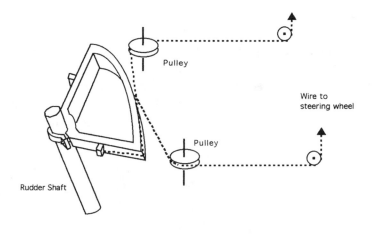

Figure 10.4 Quadrant wire steering.

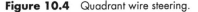

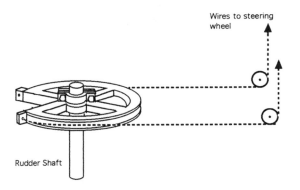

Figure 10.5 Disk drive wire steering.

wires run over a single groove in the disk's circumference and straight to the pedestal pulleys. The elimination of one set of pulleys means less friction, less possibility of loose play in the system, and fewer things that can go wrong. However, in order to work, the rudder shaft needs to be vertical. To turn the rudder the correct way the wires must be crossed in the pedestal when using the disk system.

Mechanical Steering

Mechanical steering is most easily installed in boats with aft-sloping rudder posts. This allows for a straight shaft from the center of the steering wheel to the rudder post, and only one set of gears is required (Figure 10.6). If a mechanical linkage has to go down a pedestal and then change direction 90 degrees to head aft, the extra gearing required to do this in-

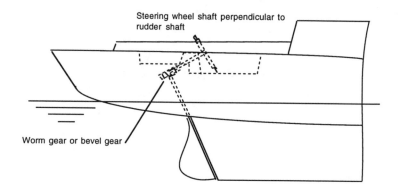

Figure 10.6 Geared steering.

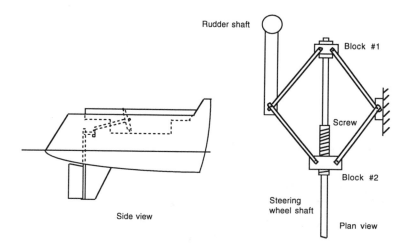

Figure 10.7 Pantograph mechanism.

creases the friction losses and loose play, making it a less attractive system. Figure 10.6 shows a straight mechanical linkage with either a set of bevel gears or a worm gear between the steering wheel shaft and the rudder stock.

Figure 10.7 shows an ingenious mechanical pantograph system. In this system, block #1 is loosely clamped between two nuts that are pinned to the shaft, and the shaft and nuts are free to turn in the block. As the steering wheel is turned, the screw drives block #2 up or down, which then causes the arms to push or pull against the arm welded to the rudder shaft. The pantograph could be made to work twice as fast by having another screw with the opposite thread going through block #1. However, the greater speed would require greater force.

An inexpensive method of mechanical steering would be to use an automobile steering system. This equipment is designed to change the rotating motion of the steering wheel to reciprocating motion at the front wheels—just what you need for the boat. This equipment is cheap and easily available at car wreckers. Best of all, because it is designed to be used in a hostile environment with dirt and water thrown up from the road, the gearing mechanism is totally enclosed. (I've never tried this but it should work.)

Hydraulic Steering

The third type of wheel control of rudders is exerted through a hydraulic system. For anyone wondering how reliable a hydraulic system can be, just think about an automobile's brake system. All cars use hydraulics for this crucial safety system. If a car won't go, it's a nuisance—if it won't stop, however, it's a disaster. Hydraulic brakes operate under the car where they are exposed to flying stones, water, ice, and salt, and yet they last for years with no maintenance required.

Figures 10.8 and 10.9 show a Vetus hydraulic system used on a powerboat. The white can-shaped object in front of the steering wheel in Figure 10.8 is the transmitter; two

Figure 10.8   Hydraulic system.

Figure 10.9   Double-acting hydraulic cylinder.

hydraulic lines run aft from there to the double-acting hy-draulic cylinder shown in Figure 10.9. This cylinder is pinned at one end to a strong bracket that is bolted to a frame welded to the hull. The piston of this cylinder is attached to an arm secured to the rudder shaft. Extra piping can be led from a junction block to a secondary steering position on the flying bridge.

Hydraulic steering can be used with any type of rud-der, even an outboard rudder if a bit of ingenuity is used. The advantages of a hydraulic system are:

- no friction in gears or pulleys,
- no loose play like that in wire or mechanical systems,
- no corrosion problems,
- it is no more expensive than a wire system,
- it is very reliable, and
- primary and secondary steering can be located anywhere.

*This Pedro 1000, a 10-meter powerboat, shows
a well-thought-out stern arrangement.*

The swimming platform provides an extension to the flat bottom, and the sturdy ladder will give easy access from the large after-deck to a dinghy in the water. This boat is used to illustrate most of this chapter.

11

Engines

For the past 100 years, automobile companies have spent vast amounts of money developing gasoline engines for cars, and marine gasoline engines have benefited from this expertise. Small, reliable gasoline engines have been available for boats for many years. However, gasoline engines have two disadvantages: a highly flammable fuel and an electric ignition system that does not work well when wet.

In the past, diesel engines were large, heavy, and low powered compared to gasoline engines. But during the past 20 years, the small high-speed marine diesel engine has been developed to such an extent that most boatbuilders are now using diesel engines for everything except high-power speedboats. Due to higher engine speeds and smaller and lighter parts, diesel engines are now small and light enough to fit into the same sized space as comparable gasoline engines.

In comparing gasoline and diesel engines, remember that diesel engines are usually rated at continuous output, whereas gasoline engines are rated at maximum output. The maximum output rating is useful for automobile engines, because the maximum output is required only for very short periods of time, and normal power output is quite low. However, a boat engine will be required to run at nearly its maximum output for long periods of time, therefore its "continuous" rating is the critical factor.

How much power do you need? For an auxiliary engine in a sailboat, 3 to 4 HP per ton is sufficient. For a displacement powerboat, 6 to 10 HP per ton is plenty. For a planing boat, a much larger engine will be required, and the size will depend on whether you want to go fast or very fast. Sailboats with small "sailing" propellers will be able to deliver only a limited amount of power to the water, no matter how large their engine.

How fast can you expect to go? The commonly accepted "hull speed" of a boat, in knots, is 1.34 times the square root of its load waterline length in feet, or

$$\text{speed} \quad = \quad 1.34 \sqrt{\text{L.W.L.}}$$

This is the theoretical speed at which the waves created by the boat moving through the water are the same length as the boat. At higher speeds, the stern moves ahead of the second crest and the boat is, in effect, going "uphill" all the time. Of course, there is no sudden change in power requirement at this speed, but after this, the stern sinks more into the hollow of the wave as the speed increases. Larger amounts of power are required for each unit increase in speed.

When choosing an engine manufacturer, you would be wise to stay with the better known names, so that you can get spare parts, even in isolated areas. Names such as Volvo, Westerbeke, Yanmar, and Perkins immediately come to mind. Keep your eyes open at the local marinas to see which companies have outlets in your area, and talk to local boat owners about any problems they have had getting spare parts for their engines.

After you have decided on a manufacturer, make sure you choose an engine that is widely used in boats. A while ago we came across a boat we had been traveling with a few months earlier. The owner was having trouble getting parts for his engine, one from a well-known firm, the same as ours, but instead of installing the manufacturer's regular marine engine, he had used a heavy-duty version not normally used on boats. The engine was very well built and the bent connecting rod was beautifully made; however, there was not a single spare available in North America, and our friend was tied up for more than a month waiting for delivery from overseas. So choose a manufacturer with a worldwide support system, then choose one of his most common marine engines, and you will avoid many expensive headaches.

In Figure 11.1, a 22-HP three-cylinder Yanmar diesel is shown installed in the author's 36-foot sailboat. Instead of being hidden under the cockpit, it has been brought forward

Figure 11.1   A 22-HP three-cylinder Yanmar diesel.

so that it is easily accessible for maintenance. A three-bladed propeller is fitted, rather than a small sailing propeller, and it delivers the engine's power to the water effectively. This engine is capable of moving the boat at 6 knots in calm weather and at about 5 knots with a bit of a chop on the water. It is a remarkably small and light engine, and it has the additional advantage of being able to be started with a hand crank, a good fallback position if the batteries run down. (Several years and many thousands of miles later, I now would prefer a 30- to 40-HP engine.)

Figure 11.2 shows a 10-meter powerboat with a 60-HP four-cylinder Volvo diesel. We use this engine in the next few photographs as an example to illustrate several basic requirements of engine installation.

The two deck supports immediately above the engine are removable, so that when the three sections of deck and these two supports are removed, there is an opening large enough to lift the engine out, if necessary. The engine is se-

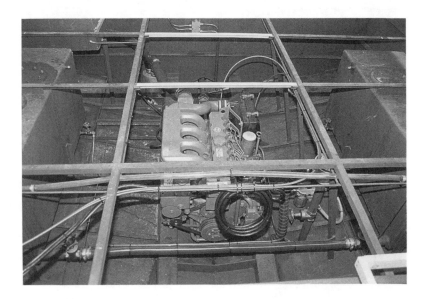

Figure 11.2   A 60-HP four-cylinder Volvo diesel.

curely mounted on the hollow box beams, which in turn are welded to the hull plating and the hollow keel. In Figure 11.2, the long fuel tanks on either side of the boat are visible. They are joined by piping at their bottoms and tops to allow the amount of fuel in each tank to be equalized. The lower cross-connect pipes are large to allow rapid equalization of the tanks when fueling. Plastic pipes are rather fragile for fuel lines so the isolation valves should be shut when fueling is finished.

The engine's cooling water intake and filter can be seen in Figure 11.3. The filter is mounted above the waterline on a steel tube leading up from the hull valve. The suction line from the stern tube can be seen entering the pipe, from the left, between the hull valve and the weed filter. The advantage of the welded pipe method of attaching hull valves can be seen in this installation—it allows the valve to be well clear of the hull and easily accessible. The filter is located above the waterline to prevent flooding should it leak, and to allow it to be cleaned even with the hull valve open.

Figure 11.3 The engine's cooling water intake and filter.

The flexible rubber mounts can be seen securely bolted to the hollow box beams (Figure 11.3). These beams and the hollow box keel are going to be sources of corrosion in this steel boat. They both have small access holes cut in them, which will allow moisture and air to enter and rust the inside of the metal, but which are too small to allow the owner to get a hand inside to prepare and paint the metal. Long hollow spaces like these beams should be airtight; then when the oxygen inside the beam is used up, rusting will stop.

Figure 11.3 also shows, on this side of the water filter, the fuel filter and water separator, with copper piping leading to the fuel tanks and to the engine. The last part of the fuel line to the engine should be of flexible tubing because the engine will move a fair amount on its flexible mounts, and solid piping will leak or fracture over time.

Figure 11.4 shows the same area from a different viewpoint. Several weaknesses can be observed here. First, items are welded onto the skin of the boat, rather than onto the reinforcing members; this is evident in the fuel filter bracket, but it also occurs with the heavy fuel tank. Two of its feet are welded to one of the frames but the other two are simply welded to the thin skin of the boat. As these tanks are likely to weigh about a quarter of a ton when full, that is a lot of weight to vibrate against a weld on a piece of ⅛-inch-thick sheet metal. A second poor practice is the use of a gate valve, rather than a ball-cock, for the hull valve for the water strainer. Looking at this valve, you cannot tell whether it is open or closed. It easily jams and takes several turns to close.

Note also in Figure 11.4 the piping supports. These are metal rods that have been welded onto the sole framing and to which the piping is strapped. This is a good way to support the weight of piping and is often omitted in low-cost or home-finished boats. The result of such an omission is that, after many hours of the heavy vibrations produced in an engine room, piping joints start to leak or the piping breaks.

Figure 11.5 shows the driveshaft with a flexible coupling at the left bolted onto the flange on the shaft coming out of the gearbox. This coupling isolates the engine vibration from

Figure 11.4 General view of engine.

the driveshaft and the hull, and the propeller vibration from the engine. It also allows for slight misalignment between the engine and the driveshaft. The thrust of the propeller is transmitted along the shaft, through the coupling, and to the boat through the engine mounts. So this flexible coupling must be designed to take thrust along the shaft as well as torque from the engine.

The right-hand end of the driveshaft disappears into the stern tube through a water seal. These water seals are available in many degrees of complexity, the simplest being a gland containing soft cord-like material (packing) wrapped around the shaft. By tightening the nut on the gland, the packing is compressed against the shaft and water is kept out. The more complicated seals have one fixture fastened to the shaft and another fastened to the stern tube, with spring-loaded faces running together.

The installation shown in Figure 11.5 has a water take-off just astern of the seal. The water is sucked out of the gland

Figure 11.5   Driveshaft.

and led into the engine cooling water intake. This will cool
the bearing and prevent leakage while the engine is running.
Note the good workmanship of using an isolation valve even
on this small line.

Under the driveshaft a hole that has been cut in the shell
plating can be seen (Figure 11.5). This hole gives access to
the hollow keel that runs the length of the boat—a potential
rust disaster area. One method of protecting the inside of the
keel would be to fill it with tar. Of course, the best solution is
not to cut the hole in the first place.

WATER-COOLED EXHAUST

The engine cooling water is drawn through a strainer, passed
through the engine block, then mixed with exhaust gases and
piped out through the transom. This simple and effective
water-cooled exhaust system is widely used because it elimi-

nates the danger of hot exhaust pipes inside the boat and reduces the noise of the exhaust. The exhaust gas and cooling water mix and pass into the water-lift canister shown in Figure 11.6. The inlet pipe on the left stops just inside the stainless steel, or fiberglass, canister, and the outlet pipe on the right goes almost to the bottom. As the exhaust gas pressure builds, the water is blown out of the exhaust pipe.

The canister must be large enough to contain all of the water in the exhaust pipe, and the pipe must slope from the engine down to the canister, otherwise the water will run back into the engine cylinders when the engine is shut down. Also, if the engine is below the waterline, there must be a siphon-breaker in the cooling water line, between the engine block and the exhaust system, to prevent engine cooling water from siphoning into the exhaust system and backing up into the cylinders.

Where the exhaust pipe exits the transom it needs to have a loop in it well above the outside water level, to keep following waves from flooding into the exhaust.

Figure 11.6 Water-cooled exhaust.

Under the exhaust pipe, at the left in Figure 11.6, you can see the rack for the batteries and the battery cable. This is not a very good location, because any leakage of water from the exhaust system might short out the batteries.

STERN TUBE

The propeller driveshaft exits from the stern of the boat via the stern tube. This tube has two main functions. The first is to support the weight of the driveshaft and propeller, and the second is to keep water from entering the boat. If the stern tube has to project a long distance from the hull to enable the propeller to clear the hull, as happens in boats with a fairly flat bottom, a bracket will be needed to support the outboard end of the tube. On larger vessels, this bracket often consists of two diagonal struts, like an inverted "A" and it is usually called an A bracket. If the outside shaft is long enough, the A bracket is separated from the stern tube and carries a separate bushing to support the propeller end of the shaft.

The bearing at the outboard end of the stern tube is usually made of a hard black material called *cutlass rubber* or of plastic. This bearing is often sold in a bronze tube, which is pressed into the stern tube. For metal boats, it is better if you can obtain the bushing without the bronze casing to avoid another source of corrosion. Press the bushing directly into your steel or aluminum stern tube.

The inboard end of the stern tube is fitted with a gland to keep the water out. This is usually a stuffing box with a couple of turns of packing wrapped around the shaft and held in place with a large hollow nut. Figures 11.7 and 11.8 show a typical small boat arrangement. The gland is attached to the stern tube by means of a thick rubber hose clamped in place. This isolates vibration from the hull and allows for slight misalignment. The shaft arrangement shown in Figure 11.8 is going to be installed in a fiberglass boat. At the

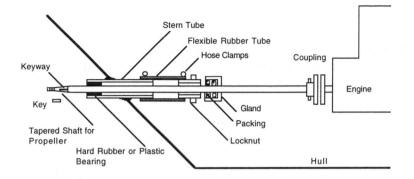

Figure 11.7 Diagram of propeller shaft and stern tube.

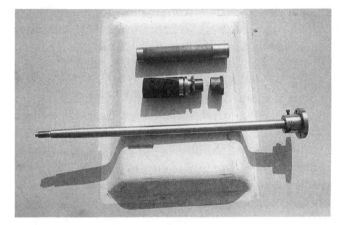

Figure 11.8 Propeller shaft and stern tube.

top is the bronze stern tube, which has been knurled to ensure a mechanical lock with the fiberglass. In our case, this tube will be the same material as the hull and will be welded in place. Inside this tube is a plain bearing made of plastic. The second item is the rubber tube that goes over the end of the stern tube, with the gland containing the packing to the right of it.

FUEL SYSTEMS

To keep a diesel engine running, the main requirement is a clean supply of fuel with no air in it. Any dirt will damage the high-pressure injection pumps, and any air will form a cushion in those pumps and prevent the fuel from being forced into the cylinder. If there is any doubt about the cleanliness of the fuel, it should be passed through a screen before being put into the tank. A nylon stocking makes a simple and easily available screen for this purpose. When the fuel is drawn from the tank, it will pass through a water separator and fuel strainer (often in the same unit), before being passed by the engine's fuel pump to the fuel filter (Figure 11.9). From there, it goes to the injection pumps and to the fuel injectors on each cylinder. Overflow from the injectors is then returned to the fuel tank. Sources of air problems are usually the joints between the tank and the fuel pump: air can be sucked in through these joints and be difficult to detect.

Diesel fuel has the drawback of providing a home for certain microorganisms that thrive in a combination of diesel fuel, moisture, and warmth. Given time, they can multiply to such an extent that they block the filter and cause corrosion in the system. The access plate in the fuel tank makes it possible to get at the inside of the tank to clean it if it is contaminated by these little creatures or by dirty fuel.

A gasoline fuel system (Figure 11.10) is simpler than a diesel system because the need for clean fuel is not as critical, and a small amount of air in the system causes no problem. The main disadvantage of gasoline is its highly flammable and explosive nature. Great care must be taken to ensure that there are no leaks in the system. A forced-air ventilation system is necessary in the engine and fuel area. The exhaust fan should have its hose leading to the bottom of the engine space, so that any fumes that gather there will be removed. Run this exhaust fan for at least five minutes before the engine is started.

Fuel tanks should be pressure-tested to 10 to 15 psi to make sure that they do not leak. Because most people do not

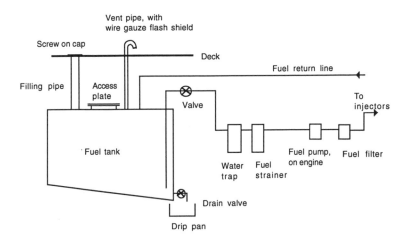

Figure 11.9 Diesel fuel system.

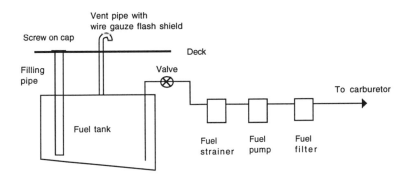

Figure 11.10 Gasoline fuel system.

have the equipment to pressure-test tanks, a simple method is to use a hose attached to one of the pipes going into the tank with the other holes sealed—a garden hose will do. Because there is about ½-psi pressure for every foot of water, a hose 20 feet high will give you a 10-psi water pressure test. Note that it has to be 20 feet high, not long, so you will find yourself leaning out of a second-or third-story window, with the tank on the ground, in order to run this test.

You can save a lot of traveling time by having the hull close to your home for the finishing work. Even if it takes only half an hour each way between your home and the work site, that's an hour a day, seven hours a week, or the equivalent of a full day's work wasted every week. This boat is being lifted into the owner's back garden so that the interior work can be done close to workshop, tools, coffee, and rest.

CHAPTER

12

Finishing the Interior

The hull is finished, the keel has been poured, holes have been cut for the hatches and portlights, sandblasting and painting are finished, and now you are ready to start the inside work. Congratulations, you are about half-finished and can now celebrate moving into the second phase of your project. Probably the work to date has been done in a builder's yard or marina, so that the dirt and noise would not disturb the neighbors. The heavy, dirty work is now finished and you should consider moving the boat closer to home. The two main requirements for boatbuilding are time and money, so if you can move the boat into your own backyard you will start to save a lot of time. It's much easier to step out the back door for the evening's work than to drive across town, especially on a snowy night.

Your first job is to mount the hatches and portlights; this will give you a weatherproof boat to work in and there will be no need for a shed or a temporary wood and plastic eyesore. Put a heater in and you can ignore the rain and snow.

WOOD

Plywood is used in many areas of the boat's interior, from heavy 14-mm plywood for the cabin sole to light 6-mm plywood for dividers in cupboards. The two main grades of plywood are exterior and interior grade. Exterior grade is the one needed in boat work because it is made with waterproof glue. There is also a "marine" plywood that is similar to the exterior grade, except that the core plies have fewer knots and holes in them, making the plywood stronger. Marine plywood would normally be used where the plywood is to be a structural part of the hull, and especially if the whole hull is to be built of plywood. But for interior finishing, it is not worth the extra cost, and the difficulty of finding a supply.

Size		Typical Uses
14 mm	(⅝ in.)	Cabin sole
11 mm	(½ in.)	Seat and berth tops
8 mm	(⅜ in.)	Lockers, interiors, bottoms of seat lockers
6 mm	(¼ in.)	Cupboards and dividers

A commonly used plywood is described as *good-one-side* or G1S. This has one side with all knotholes plugged and the surface smooth and ready for finishing; the reverse side is left rough with knotholes. Make sure you buy plywood that has had its holes plugged with wood and not with plastic wood filler—the difference becomes very obvious when it is varnished. For surfaces that will be exposed, plywood with a thin veneer of fine wood, such as teak or mahogany, can be used. (In mass-produced boats, a teak interior is plywood with a very thin veneer of teak, similar to that on so-called "teak" furniture.)

The choice of woods to be used in finishing the inside of a boat is rather like an artist's palette (Table 12.1). It is best to limit the number of types of wood visible—too many will give a hodgepodge effect. If two types of wood are used, one can be for the main surfaces and the other for trim and accent.

Table 12.1   Wood Used in Finishing the Interior of a 36-foot Steel Sailboat

Plywood	
14 mm	3 sheets
11 mm	9 sheets
8 mm	14 sheets
4 mm (mahogany)	1 sheet
Oak	10 board feet*
	12 sq ft, ⅜ inch thick
Pine	180 board feet
Cedar	670 sq ft
	Select, tongue and grooved, 4 in. wide, ⅝ in. thick
Mahogany	108 board feet
Teak	24 board feet
Plus two wood doors and four teak grab rails	

*A board foot is 144 in.³ of wood, or a piece 12 in. × 2 in., 1 in. thick.

The accepted boatbuilding woods, when boats were structurally built out of wood, had to have the properties of strength and resistance to rot. However, when it comes to finishing the inside of a metal hull, the strength is in the hull, and the wood can be chosen for looks, price, availability, and workability. Cedar, for example, is a useful finishing wood, because it is light and rot-resistant, smells nice, is easy to cut and shape, comes in warm colors, and is not too expensive. It is, however, weak and soft, and therefore easily broken, scratched, or dented. If you are going to use cedar for the inside of the boat, you will have to use a stronger wood for framing and furring strips. For great strength, oak could be used, but it is expensive and hard to work. For reasonable strength at a reasonable price, pine is a better choice. Pine is also popular for interior surfaces, because it is strong and light in color. For trim, mahogany and teak are popular. They are easy to work with, virtually impervious to rot, and, most of all, are beautiful when finished.

For wood trim on the outside of the boat, teak and mahogany are the popular choices. Teak is especially popular because it can be left alone to weather to a pleasant gray color—its oils will protect it from decay. Or if you have the time, you can finish it with various products to maintain its original color. Oiled teak darkens fairly rapidly, so it is worthwhile to look for more advanced plastic finishes if you want the teak to retain its natural color. Mahogany is usually varnished to bring out its rich colors, a time-consuming annual job.

Any wood used in your boat must be dried—kiln drying is the most common method—or you will find it warping and splitting after a year or two.

INSULATION

Insulation protects against heat and cold, keeping the heat in on cold days and keeping it out on hot days. It also deadens noise: without insulation, living in a metal boat would be like living inside an oil drum, with the sound of every wave being magnified.

In insulating a house, a vapor barrier is placed on the warm side of the wall. This prevents moisture from condensing in the insulation; the warm damp air permeates the insulation until it reaches its condensation temperature. Without the vapor barrier, the insulation would quickly become soaking wet and useless. As it is impractical to create a vapor barrier in a boat, the use of fiberglass insulation is not satisfactory. The insulation must act as its own vapor barrier as well as a heat and noise insulator.

Two common methods of insulating are with spray-on foam plastic (polyurethane) and stick-on Styrofoam (Dow

Chemical trademark for polystyrene). The sprayed-on insulation, the same as that used on vans and campers, has the advantage of covering all of the metal hull, including the stringers and frames; it has the disadvantage of being expensive and messy.

If you opt for Styrofoam, the one to use is the high-density blue Styrofoam SM, which is normally used for insulating the exterior walls of basements and buildings. This rigid extruded polystyrene is easy to cut with a sharp knife, and can be stuck to the hull with normal panel adhesive, the type that comes in tubes and is applied with a caulking gun. The adhesive should be put on as a bead completely around the back of the section being stuck on to prevent moisture getting behind the section and condensing on the metal hull. Do not try to use contact cement because it will dissolve the Styrofoam. A 1-inch layer of this material has an R5 insulation factor and will fit under 1½-inch angle iron stringers and ribs, as shown in Figure 12.1.

Figure 12.1 Styrofoam insulation.

Figure 12.2 Sprayed-on insulation.

Figure 12.2 shows an example of polyurethane foam insulation that has been sprayed in place over the entire inside of the hull. The builder insulated before bolting wood strips to the metal frame. He will now have the messy job of cutting out the foam so that he can bolt on his wood attachment strips. He has predrilled the frames (one hole can be seen) to facilitate bolting on the wood. The task of drilling holes in flat-stock frames and stringers is fairly easy before the hull plates are attached, but awkward afterward because it is difficult to get a normal hand drill close enough to the skin to drill holes straight through the frames. A simple type of scuttle arrangement is shown, with a sheet of transparent plastic bolted onto the cabin side.

WATER TANKS

These can be made of various materials. In an aluminum boat, aluminum would be the first choice for water tanks and holding tanks (for the head); for steel boats, stainless steel, fiberglass, or ready-made plastic tanks can be used. Built-in steel tanks, using the hull as part of the tank, are not advisable, as the hull could rust inside the tank without your realizing it.

Figure 12.3 shows the construction of a typical tank made of rigid sheets of fiberglass. These sheets can be worked like thin sheets of plywood. They can be cut on a table saw, then stuck together with several layers of fiberglass mat soaked in polyester resin layered up on their corners. Such preformed sheets have a layer of wax on them that needs to be sanded off before the resin will hold. (It took me a week of trial and error to learn this simple fact.)

Figure 12.3 shows the three pipes needed in all tanks: the suction pipe extending to the bottom of the tank; the filling pipe; and a small exhaust pipe to let air out of the tank as it is being filled. In this case, the tank shown is the holding tank for the head. The exhaust pipe was led outside of the cabin. Figure 12.4 shows two freshwater tanks in place in the

Figure 12.3 Holding tank.

Figure 12.4 Water tanks.

keel, with clean-out access plates visible. Foam plastic was extruded into the space between the tank and the keel to hold the tank securely in place. The keel is a good location for tanks because 100 gallons of water weigh about half a ton, and, when full, the tanks increase the boat's stability. This location does not change the boat's trim as the water is used.

PLUMBING

Plastic plumbing equipment is ideal for use in a metal boat because there is no corrosion problem and no chance of electrolysis between the piping and the hull. The piping can simply be cut to length and stuck together with the appropriate adhesive. Each type of plastic has its own glue, but all are easy to use.

Figure 12.5 shows a two-tank freshwater arrangement. Each tank will need its own valve in the fill line and the suc-

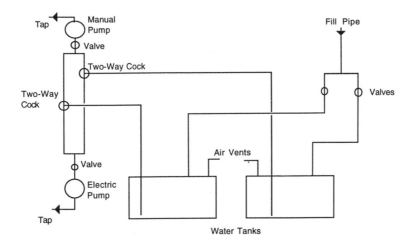

Figure 12.5   Freshwater system.

tion line. When the tanks arc full, water will start to come out of the air vent, so this should be located where the overflow will do no damage. Separate valves are installed in each suction line, so that a choice can be made as to which tank to use. If one tank is completely emptied before the next is started, you will know how much water you have used and how much is left. Emptying each tank in turn also prevents water from becoming stagnant as a result of staying in the tank too long.

To lift the water to the tap, an electric or a manual pump can be used. Pumps using a rotary impeller are not much use for tanks located in the keel, because they require a flooded suction and therefore must be installed lower than the tanks from which they draw. Manual pumps are usually of the reciprocating type that will easily suck water up a few feet. A well-built brass or bronze pump looks tiddly and will give reliable service for many years, but you can only wash one hand at a time. Foot pumps are the most convenient because they leave both hands free to wash, and have no electrical parts to corrode and short out.

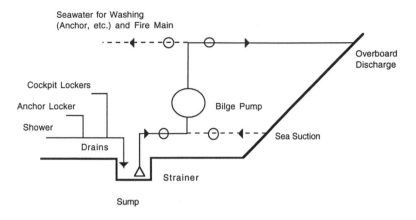

Figure 12.6   Seawater and drain systems.

The second water system on the boat is the drain system (Figure 12.6). Lockers that open to the outside, showers, and any other source of water should have drainpipes from them to the lowest point in the keel or bilge. A bilge pump then pumps the water overboard. A well-designed drain system will help ensure that you have a dry boat and avoid mildew.

The bilge pump can also be used to provide water for a fire hose. For this, the pump needs two suctions, one from the bilge and one from a sea connection, each having a separate valve, and two discharges, one overboard and one to the fire main, each with its own separate valve. Then the bilge pump can pump either from the bilge overboard, or from the sea to the fire hose, which is also useful for washing down the anchor chain. In freshwater lakes, this arrangement will provide an upperdeck shower for the whole crew at once.

LOCKERS

Figure 12.7 shows a cockpit locker being installed (view starboard quarter facing aft). The construction method consists

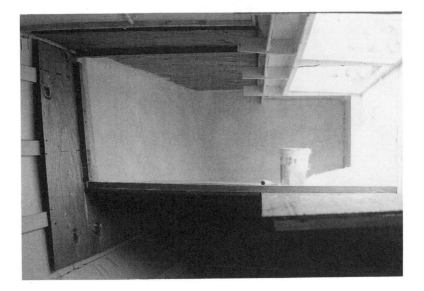

Figure 12.7 Locker construction.

of building the shape of plywood, then lining the inside with fiberglass to produce a waterproof, rot-resistant container that is easily cleaned. The hull is insulated, then the plywood is attached to the stringers and deck frames, using stainless steel pop rivets or furring strips, as in the rest of the boat.

Looking forward in Figure 12.8 you see two anchor lockers. The forward locker has a hatch leading to the foredeck, and contains a kedge anchor and 200 feet of nylon line, plus other deck gear. The second, lower, plywood container is the chain locker for the main anchor rode. Because 200 feet of ⅜-inch chain weigh about 300 pounds, this locker has to be strongly built and as low as possible.

The two lockers of Figure 12.8 were built with their sides attached to the stringers, using stainless steel pop rivets. All lockers exposed to the weather need to have drains in the bottom. These drains should be led back to a common sump in the bilge, so that the drainage can be easily pumped overboard. (The telephone seen in the photo proved to be a useful addition to the boatbuilding environment.)

Figure 12.8 Anchor lockers.

To attach the inside of the boat to the hull, vertical strips of wood are bolted onto the stringers, then the internal wood-work can be screwed onto these strips. In house building, these are called *furring* strips, and this seems to be a suitable name for them in boats also. If you are going to use spray-on foam insulation, then the furring strips should all be bolted on before the foam is sprayed on.

In Figure 12.9, some of the furring strips, of 1-inch × 1½-inch oak, have been bolted onto the stringers, and the builder (me) is welding on some brackets that he should have thought of before the insulation was put in.

The frame for the forward double berth is installed (Figure 12.10). The open lattice of pine strips ensures free movement of air behind the boat's interior lining to prevent moisture and mildew. The wood slats are screwed onto the oak furring strips, and the cross-pieces at the top are bolted onto brackets welded onto the stringers.

Figure 12.9 Forward V-berth.

Figure 12.10 Forward V-berth.

The forward V-berth cabin is starting to take shape in Figure 12.11. The tongue-and-groove cedar planks are glued and screwed onto the furring strips. A gap is left between the top of the hull planking and the deckhead to allow for movement of air between the hull and the cedar lining. As the hull is heated by the sun, the air between the hull and the interior lining is heated and rises, drawing in air from between the slats at the bottom and expelling it through the gap at the top. During the night, the hull cools and the movement of air is in the other direction. This ensures that the space between the hull and the interior lining is kept well ventilated and free from condensation and mildew.

After the hull is planked, the forward bulkhead is built. A door will be cut in this bulkhead later to give access to the chain locker. The bulkhead separating the heads and the hanging locker from the V-berth is shown, looking forward, in Figure 12.12. The brace is bolted to steel brackets welded to the carlin strip. The vertical planks are screwed to a curved piece fitted to the deckhead, the brace, and a deck strip. To ensure that the bulkheads line up with each other, the brace is carried across the full width of the boat. When the bulkhead is completed, the brace will be cut away at the passageway. The inlet from a Dorade-style ventilator can be seen on the left of Figure 12.12. The gap between the inlet pipe and the deckhead liner allows for movement of air behind the liner. This hole will later be trimmed with a mahogany ring.

Looking aft into the starboard quarterberth (Figure 12.13), the same method of construction and ventilation can be seen. The cockpit lockers shown earlier have now been completed and faced with cedar planks simply nailed (using bronze nails) to the plywood lockers. The boat's batteries will be located under the quarterberth, so large holes have been cut into the bottom of the below-berth locker for ventilation and tie-down straps, plus ventilation holes into the adjacent engine compartment, which has its own exhaust fan. The seat framing is bolted to steel brackets welded to the stringers.

Figure 12.11 The V-berth.

Figure 12.12 A bulkhead.

Figure 12.13 Starboard quarterberth.

Figure 12.14 Port cockpit locker, facing aft.

Figure 12.14 gives us a view of the port cockpit locker, facing aft. This is going to be a large locker, big enough to climb into to give easy access to the fuel tank, and to provide storage space for a couple of folding bicycles and spare fenders and what have you. Heavier sheets of plywood are used here, so they can carry a person's weight. They are screwed onto the furring strips, which are bolted onto the stringers. The plywood liner gives a finished appearance in the locker, as well as a place to hang things. It also allows you to stick insulation on the hull, which keeps the boat much quieter. Many people, however, simply leave the lockers bare metal, which is much simpler. On the left you can see the engine controls and the top of the mild steel fuel tank. The black plastic pipes coming from the lockers are drains that lead down to the bilge sump pump.

The berth shown in Figure 12.15 illustrates a typical method of construction. The back of the seat is framed, using 1-inch × 1½-inch pine, glued and screwed to the surrounding liner and bulkheads. For shelves, lighter ½-inch × ½-inch stock is suitable. Plywood is then fitted to the frames. The openings in the seat giving access to the below-seat lockers should be as large as possible. Those shown proved to be too small for convenience, and were later enlarged to the full size of the seat bottom.

Shelves in the dinette area start out with ½-inch × ½-inch stock fastened to the liner and bulkheads, plus a central support (Figure 12.16). The shelves are fitted, and ½-inch × ½-inch stock is glued and screwed to the front edge, so that the shelves can be attached to the mahogany facing (Figure 12.17). The central support is made of heavier 1-inch × 1½-inch framing because it will carry the weight of one end of the table.

Finally, mahogany 1-inch × 6-inch planks are fastened to the shelves and side supports, and mahogany trim on the vertical surfaces (Figure 12.18). The table is removable so that it can fit between the seats and form the bottom of a double berth. The top shelf will be closed in with sliding doors to stop objects from falling out in heavy weather and to hide

Figure 12.15   Seat berth construction.

Figure 12.16   Building shelves.

Figure 12.17 Building shelves.

Figure 12.18 The dinette area takes shape.

the junk that will collect there. Below the shelves can be seen
the gaps between the slats, which allow air to circulate behind the wood to prevent condensation.

GALLEY

Arranging the galley area is one of the more interesting layout problems and one on which much advice is usually forth-
coming. Because this is one of the most used areas of the
boat, a lot of thought should go into the positioning and size
of shelves, stove, icebox, sink, stowage for pots, pans, dishes
and food, etc. The first few attempts are most easily done on
paper. Then, when you are fairly sure of what you want, try
a full-scale mock-up with strips of wood to outline the
counters and cupboards. Then see whether you can reach
everything, and if you have a smaller sailing companion, see
whether that person too can reach all the corners without
risking getting burned on the stove.

In Figure 12.19, a cardboard box has been folded into
the exact measurements of the stove. It's much easier to move
a cardboard box weighing a few ounces than a
counter-balanced stove weighing a couple of hundred
pounds. The counters have been laid out, using long pieces
of wood. Check the counter heights and depths: Is there room
to turn around? Reach under the sink? Will the stove swing
without hitting the hull? Can you reach the shelves behind
the stove?

The area under the stove has been covered with a sheet
of copper to make it easier to clean up after the inevitable
spills (Figure 12.20). Above the stove, another sheet of copper protects the wood from the heat of the burners.

Install a full-sized sink—there's nothing more irritat-
ing than those silly little marine sinks sold at exorbitant prices
in marine stores. The full-size sink is also convenient for
cleaning crab, fish, abalone, and other gifts from the sea.

Figure 12.19 Galley.

Figure 12.20 Stove area.

The Icebox

Are you going to use a refrigerator? Are you going to build a huge icebox to carry two or three weeks' supplies? Do you need an icebox at all?

If you have a powerboat, a refrigerator is an obvious choice because there is power available to run it while you are at sea, and probably an auxiliary generator or shore power while you are at anchor or alongside. For a sailing boat, a refrigerator can be a real nuisance and you will find a lot of your lounging time goes toward keeping your batteries up for the hungry fridge.

My first thought about iceboxes was that, if you can go ashore and buy ice, then you may as well buy fresh food instead and eat it within a couple days. Because ice lasts for only a few days in warm weather, obtaining ice and toting the heavy stuff to the boat seemed more trouble than it was worth, and in many parts of the world, it isn't even available. However, as can be seen in Figure 12.21, I finally did install an icebox, deciding on a portable one mounted in a space that was further insulated with 2 inches of Styrofoam. However, I must admit that, after living aboard full time for several years, we very rarely put any ice in it. Still, it does make a clean storage space for food and for ice cubes when we have visitors over for drinks.

Figure 12.22 shows a more conventional icebox, with the inner box consisting of 2 inches of polyurethane lined with fiberglass. A track holding clear plastic with holes in it allows the water from the melting ice to run into the bottom of the box to a drain (do not drain the icebox to the bilge or else the boat will very quickly smell terrible) and keeps the food out of the water. Around this inner box is an outer one made of another 2 inches of polyurethane, which is coated on both sides with aluminum foil to improve its insulating properties. An insulated top, which will incorporate a smaller hatch, has yet to be installed.

Figure 12.21 Icebox.

Figure 12.22 Icebox.

CHART TABLE

A good-sized chart table is a must if you plan on anything more than weekend cruising. The number of charts and navigation books required for long coastal passages is staggering. Our first trip in our boat was from Ottawa, Canada, to the Bahamas. This trip necessitated 41 charts, a book of charts for the Erie Canal, plus half a dozen reference books and guides. The same number again are needed to cruise the islands of the Bahamas.

The top of the chart table should be used for navigation work only, and it should not be necessary to lift the table top to get at the charts below. The beginning of a chart table is shown in Figure 12.23. It has two shelves for charts and a large stowage area beneath the table top.

In Figure 12.24, the shelves under the table top, with a fiddle a few inches from the front, can be seen. The charts are stored in folders behind the fiddle, and all sorts of useful items can be kept in the front section. Two large shelves simplify the organization of the charts: one for the chart folio that you are using, and one for all the rest.

The completed chart table has been faced with mahogany and the table top covered with Formica (Figure 12.25). The Formica provides an easy-to-clean surface that is hard enough to resist damage from pencil and compass points and is a light-colored contrast to the wood interior. The mahogany doors have stainless steel piano hinges on their lower edges, and when open, protrude only 6 inches into the passageway. If drawers had been used, they would have blocked the passageway when open, and would have been easily damaged by heavy bodies falling against them. When closed, they would have occupied a lot of space under the chart table that more usefully holds charts. A fluorescent light provides good illumination for reading charts. A space has been left under the upper cabinet to allow charts to be spread out. The upper left space in the cabinet is destined for electronic equipment, in this case, a VHF radio and a Loran set. The right-hand space is a bookshelf.

Figure 12.23 Chart table.

Figure 12.24 Chart table.

Figure 12.25 Chart table.

Note the plywood seatback to the left of the chart table in Figure 12.25. Holes in plywood can be filled with wood or plastic wood. The difference is not very obvious when the plywood is dry, but as soon as it is varnished the plastic wood fill shows up as in the photo. Make sure you buy plywood with wood-filled holes.

PORTLIGHTS

Large opening portlights are a boon in hot weather; they are expensive but well worth it. In Figure 12.26, a wood backing, made by gluing pine planks edge to edge, was built to extend well past the edge of the portlight. The center was then cut out to accommodate the frame. Since the side of the cabin is curved, the backing needs to be cut away on the outboard side in order to fit. A good seal between the wood and the cabin side is essential to prevent leaks. A generous layer of silicone sealant is helpful here.

The inside liner was then attached to the backing which was securely bolted to the cabin side by the portlight frame. The opening was then ready for a decorative trim (Figure 12.27), in this case mahogany. More trim will be used in the angle between the deckhead and the cabin sides.

Figure 12.26 Portlights.

Figure 12.27 Portlights.

Boat Designs

On the following pages, a few examples are shown of the many types of boats that are being designed and built in steel and aluminum. Several of the designers listed in Appendix B (Ken Hankinson Associates, Bruce Roberts, Glen-L Marine, and Benford Design Group) have catalogues of their boat plans that they will forward at a nominal cost. These are great dream books for the armchair builder. Scott Sprague can be found at Accumar Corporation and Ken Hankinson Associates distributes the plans of several other designers.

Custom designs can be quite expensive, so you should take a good look at the many boat designs that are already available before you decide to go with a custom design. After all, a boat is just a floating box with pointy ends, and with the many thousands of variations on this basic theme already completed there is bound to be a design already available that will suit your needs—and even your wants.

When you find a design that looks suitable ask for a set of *study plans*—these are inexpensive drawings that will tell you all you need to know about a design in order to decide whether or not this is the boat for you.

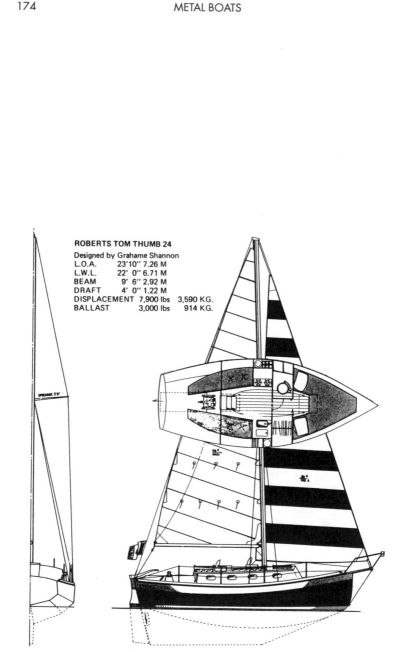

ROBERTS TOM THUMB 24
Designed by Grahame Shannon
L.O.A.   23'10'' 7.26 M
L.W.L.   22' 0'' 6.71 M
BEAM   9' 6'' 2.92 M
DRAFT   4' 0'' 1.22 M
DISPLACEMENT 7,900 lbs 3,590 KG.
BALLAST   3,000 lbs   914 KG.

Courtesy of Bruce Roberts Ltd.

Specifications

LOA	27' 8"
LWL	24' 2"
Beam	10' 1"
Disp.	13,527 lb.
Ballast	4,500 lb.
Power	10-20 HP
Sail Area	395 Sq. Ft.
Headroom	6' 4"

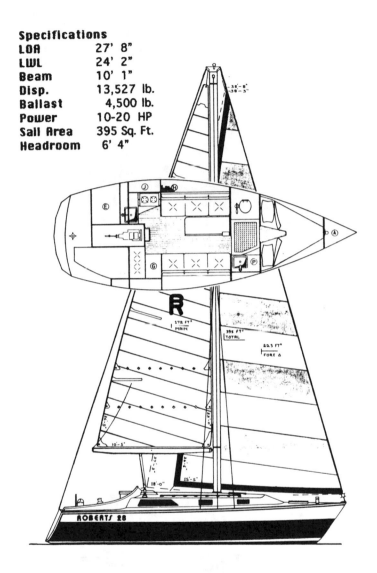

ROBERTS 28 SLOOP – STEEL

Courtesy of Bruce Roberts Ltd.

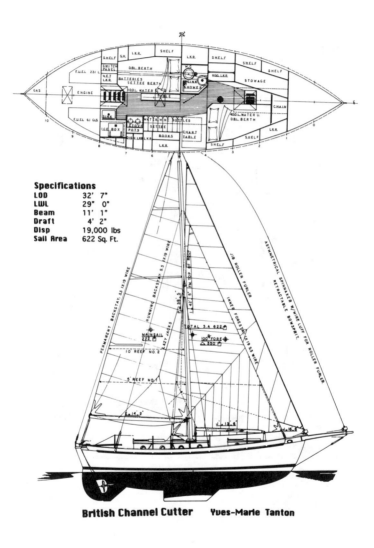

Specifications

LOD	32' 7"
LWL	29" 0"
Beam	11' 1"
Draft	4' 2"
Disp	19,000 lbs
Sail Area	622 Sq. Ft.

British Channel Cutter   Yves-Marie Tanton

Courtesy of Tanton Inc.

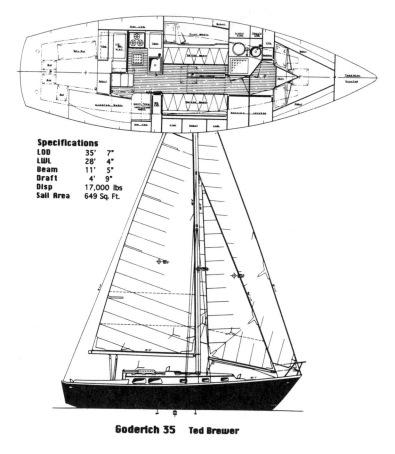

Specifications

LOD	35'	7"
LWL	28'	4"
Beam	11'	5"
Draft	4'	9"
Disp	17,000 lbs	
Sail Area	649 Sq. Ft.	

Goderich 35 Ted Brewer

Courtesy of Ted Brewer Yacht Design Ltd.

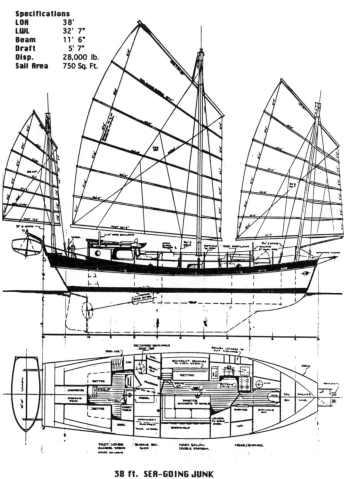

Specifications
LOA	38'
LWL	32' 7"
Beam	11' 6"
Draft	5' 7"
Disp.	28,000 lb.
Sail Area	750 Sq. Ft.

38 ft. SEA-GOING JUNK
Scott Sprague

Courtesy of Scott B. Sprague

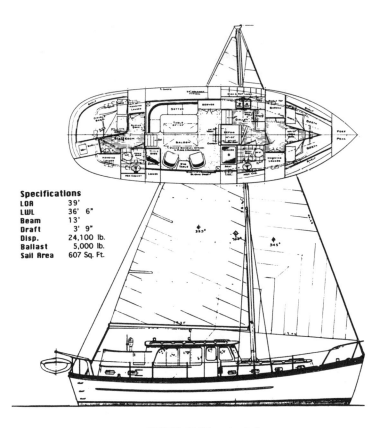

Specifications

LOA	39'
LWL	36' 6"
Beam	13'
Draft	3' 9"
Disp.	24,100 lb.
Ballast	5,000 lb.
Sail Area	607 Sq. Ft.

POWER-PLAY Motor Sailer
Charles Wittholz

Courtesy of Ken Hankinson Associates

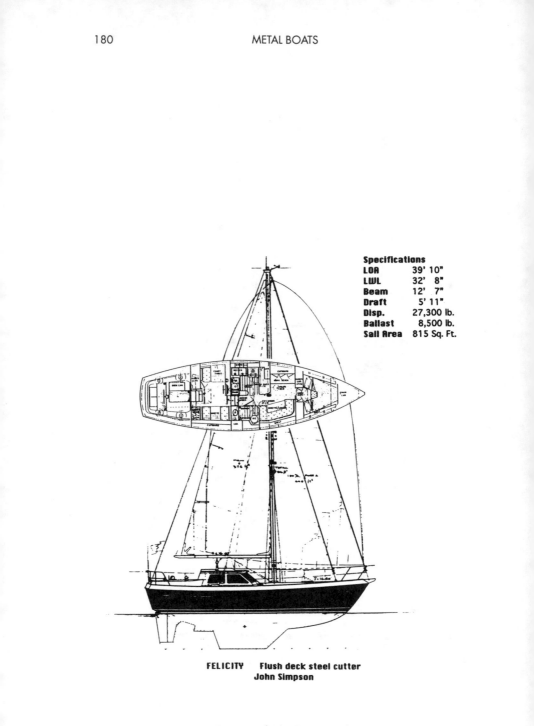

Specifications

LOA	39' 10"
LWL	32' 8"
Beam	12' 7"
Draft	5' 11"
Disp.	27,300 lb.
Ballast	8,500 lb.
Sail Area	815 Sq. Ft.

**FELICITY   Flush deck steel cutter
John Simpson**

Courtesy of John Simpson Ltd.

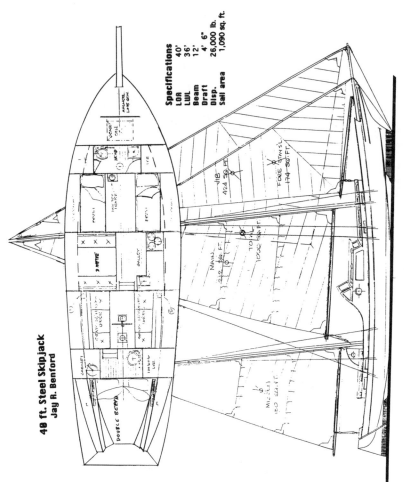

48 ft. Steel Skipjack
Jay R. Benford

Specifications

LOA	40'
LWL	36'
Beam	12'
Draft	4' 6"
Disp.	26,000 lb.
Sail area	1,090 sq. ft.

Courtesy of Jay R. Benford

LOA: 40' 6"
LWL: 33' 0"
Beam: 12' 9"
Draft: 5' 10" – 7' 0" (or 4' 9"
     centerboard)
Displacement: 24,000 lbs
Ballast: 9,000 lbs
Sail Area: 910 sq ft
SA/D: 17.5
D/L: 298
Auxiliary: Westerbeke 42-B
Fuel: 150 gals
Water: 200 gals

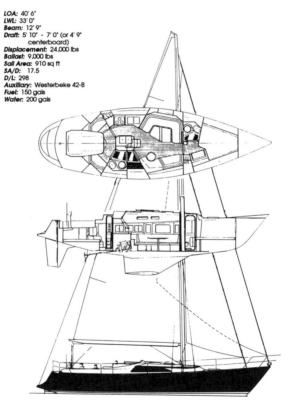

Kaufman 40
Aft Cockpit Cutter
Sail Plan

Courtesy of Kaufman Design Inc.

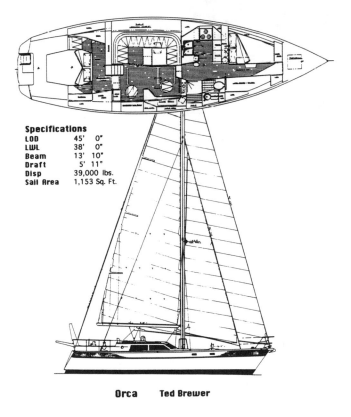

Specifications

LOD	45' 0"
LWL	38' 0"
Beam	13' 10"
Draft	5' 11"
Disp	39,000 lbs.
Sail Area	1,153 Sq. Ft.

Orca   Ted Brewer

Courtesy of Ted Brewer Yacht Design Ltd.

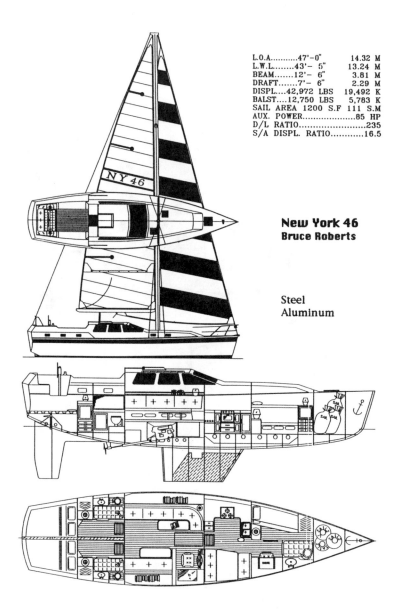

L.O.A...........47'-0"	14.32 M
L.W.L........43'- 5"	13.24 M
BEAM.......12'- 6"	3.81 M
DRAFT.......7'- 6"	2.29 M
DISPL....42,972 LBS	19,492 K
BALST....12,750 LBS	5,783 K
SAIL AREA 1200 S.F	111 S.M
AUX. POWER....................85 HP	
D/L RATIO........................235	
S/A DISPL. RATIO............16.5	

New York 46
Bruce Roberts

Steel
Aluminum

Courtesy of Bruce Roberts Ltd.

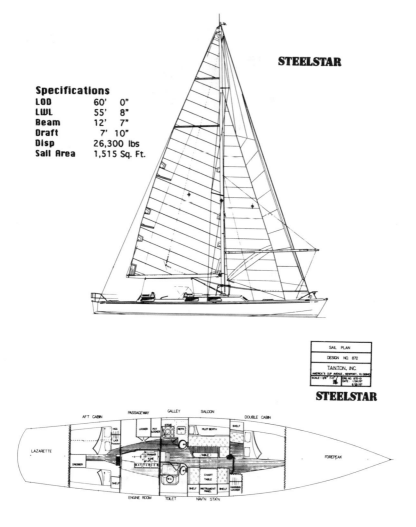

STEELSTAR

Specifications

LOD	60'	0"
LWL	55'	8"
Beam	12'	7"
Draft	7'	10"
Disp	26,300 lbs	
Sail Area	1,515 Sq. Ft.	

SAIL PLAN

DESIGN NO. 872

TANTON, INC.

AMERICA'S CUP AVENUE, NEWPORT, RI 02840

SCALE - 3/8" 1'-0"

STEELSTAR

Courtesy of Tanton Inc.

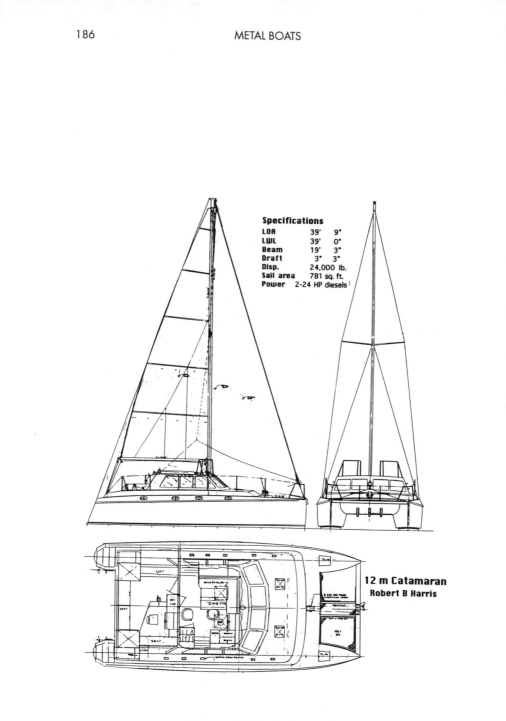

Specifications

LOA	39'	9"
LWL	39'	0"
Beam	19'	3"
Draft	3"	3"
Disp.	24,000 lb.	
Sail area	781 sq. ft.	
Power	2-24 HP diesels	

12 m Catamaran
Robert B Harris

Courtesy of Robert B. Harris Ltd.

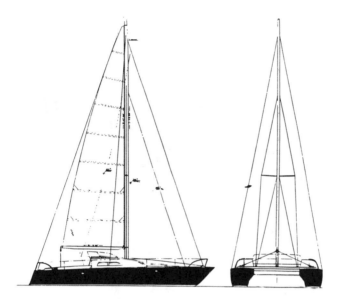

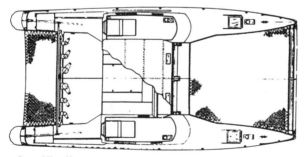

Specifications

LOA	40'	0"
LWL	35'	10"
Beam	20'	0"
Draft-board up	1'	3"
Disp. loaded	10,000 lb.	
Sail area	715 sq. ft.	
Power	2-18 HP outboards	

40 ft. Aluminum Catamaran
Robert B Harris

Courtesy of Robert B. Harris Ltd.

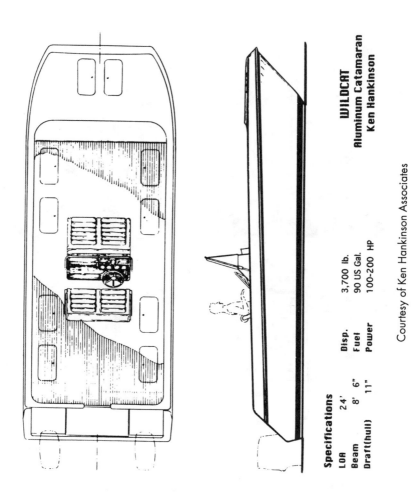

Specifications

LOA	24'	
Beam	8' 6"	
Draft(hull)	11"	

Disp.	3,700 lb.	
Fuel	90 US Gal.	
Power	100-200 HP	

WILDCAT
Aluminum Catamaran
Ken Hankinson

Courtesy of Ken Hankinson Associates

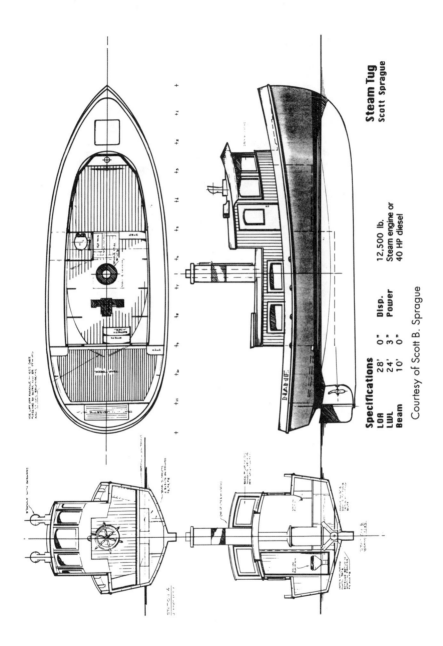

Steam Tug
Scott Sprague

Specifications

LOA	28' 0"	Disp.	12,500 lb.
LWL	24' 3"	Power	Steam engine or
Beam	10' 0"		40 HP diesel

Courtesy of Scott B. Sprague

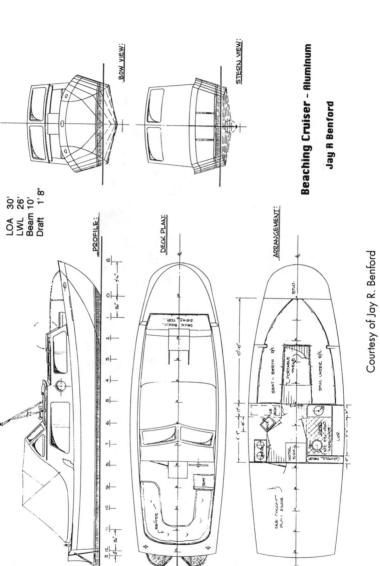

LOA 30'
LWL 26'
Beam 10'
Draft 1' 8"

BOW VIEW:

STERN VIEW:

PROFILE:

DECK PLAN:

ARRANGEMENT:

Beaching Cruiser – Aluminum

Jay R Benford

Courtesy of Jay R. Benford

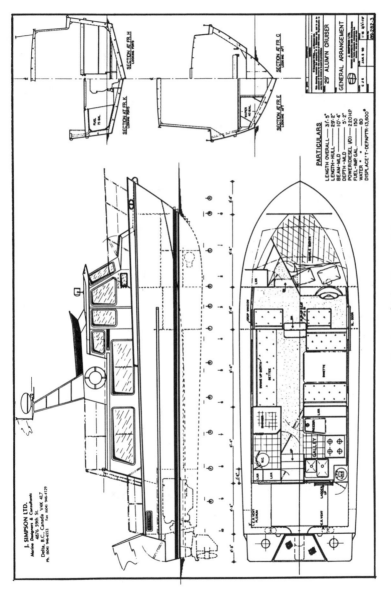

Courtesy of John Simpson Ltd.

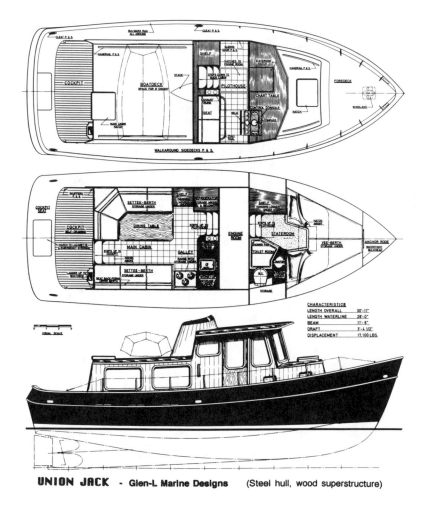

UNION JACK - **Glen-L Marine Designs** (Steel hull, wood superstructure)

Courtesy of Glen-L Marine Designs

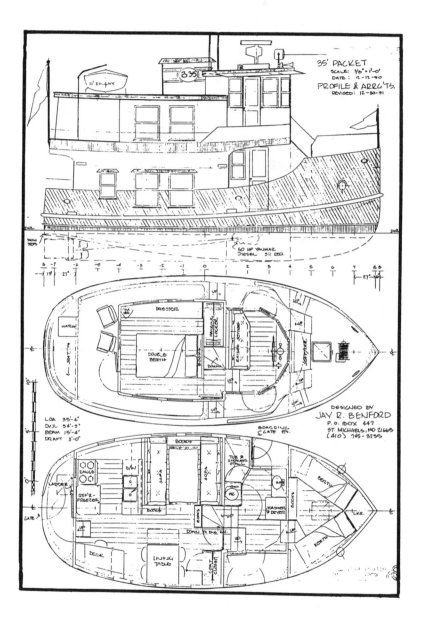

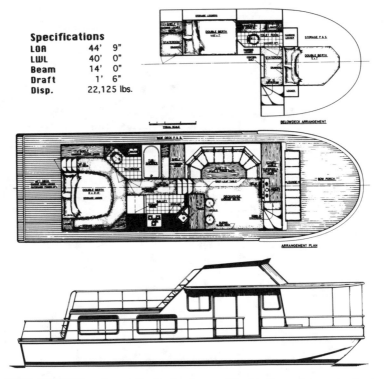

Specifications

LOA	44'	9"
LWL	40'	0"
Beam	14'	0"
Draft	1'	6"
Disp.	22,125 lbs.	

BON VOYAGE - **Glen-L Marine Designs**   (Steel hull, wood superstructure)

Courtesy of Glen-L Marine Designs

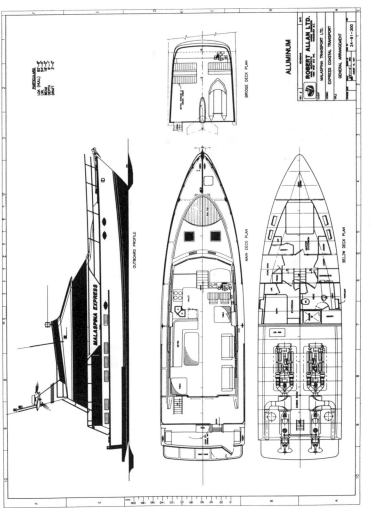

Courtesy of Robert Allan Ltd.

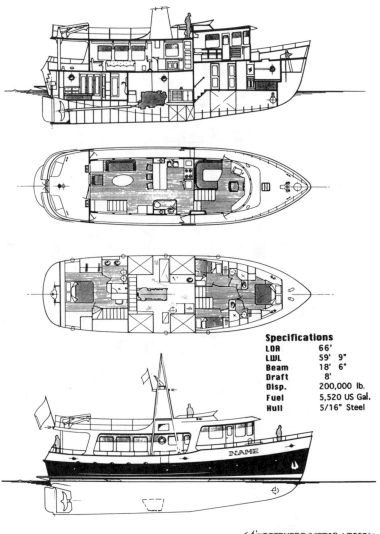

Specifications

LOA	66'
LWL	59' 9"
Beam	18' 6"
Draft	8'
Disp.	200,000 lb.
Fuel	5,520 US Gal.
Hull	5/16" Steel

64' OFFSHORE MOTOR VESSEL
Scott B. Sprague, Accumar Corp.

Courtesy of Scott B. Sprague

B

Naval Architects and Designers

Accumar Corporation
1180 NW Finnhill Road
Poulsbo, WA 98370

Robert Allan Ltd.
1690 West Second Avenue
Vancouver, BC, Canada
V6J 1H4

Jay R. Benford
605 Talbot Street
P.O. Box 477
St. Michaels, MD 21663

Ted Brewer Yacht Design Ltd.
1055 West Jordan Road
Burlington, WA 98233

George Buehler Yacht Design
Box 966
Freeland, Whidbey Island,
WA 98249

Glen-L Marine Designs
9152 Rosecrans
Box 1804/SL3
Bellflower, CA 90706

Ken Hankinson Associates
P.O. Box 2551
La Habra, CA 90631

Robert B. Harris Ltd.
611 Alexander Street, #408
Vancouver, BC, Canada V6A 1E1

Kaufman Design Inc.
222 Severn Avenue, Box 4219
Annapolis, MD 21403

Bruce Roberts
P.O. Box 1086
Severna Park, MD 21146

John Simpson Ltd.
4876 59th Street
Delta, BC, Canada

Tanton Inc.
America's Cup Avenue
P.O. Box 270
Newport, RI 02840

C

Useful Weights and Measures

1 centimeter (cm)	=	0.39 inches
1 inch (in.)	=	2.54 cm
1 foot (ft)	=	0.305 meter
1 cubic foot (ft³)	=	6.23 gallons (Imperial)
	=	7.48 gallons (U.S.)
	=	0.028 cubic meters
	=	28.3 liters
1 cable	=	608 feet
	=	0.1 nautical mile
1 cubit	=	18 inches (for building arks)
1 fathom	=	6 feet
1 foot of water	=	0.434 psi
	=	0.0295 atmospheres
	=	0.883 inch of mercury
1 gallon (Imperial)	=	10 lbs freshwater
	=	1.2 gallons U.S.
	=	160 ounces
	=	4.54 liters

1 gallon (U.S.)	=	8.34 lbs water
	=	0.834 gallons (Imperial)
	=	128 ounces
	=	3.78 liters
1 kilogram (kg)	=	2.2 lbs
1 inch of mercury	=	0.49 psi
	=	1.13 feet of water
1 kilowatt (kW)	=	1.34 HP
1 knot	=	1 nautical mile per hour
	=	101 feet per minute
	=	30.89 meters per minute
1 liter (l)	=	0.22 gallon (Imperial)
	=	0.26 gallon (U.S.)
	=	0.035 cubic feet
	=	0.88 quart (Imperial)
	=	1.06 quart (U.S.)
1 meter (m)	=	0.547 fathoms
	=	3.28 feet
	=	39.37 inches
	=	100 centimeters
1 mile (statute)	=	1760 yards
	=	5280 feet
	=	1.609 kilometer
1 mile (nautical)	=	1.15 miles (statute)
	=	2027 yards
	=	6080 feet
	=	1.85 kilometer
1 ton (British, long)	=	2240 lbs
1 ton (Canadian, U.S.)	=	2000 lbs
1 tonne (metric)	=	1000 kilograms
1 league	=	about 3 miles (statute)

Useful Formulas

Triangle

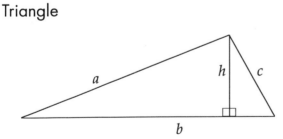

Area = $1/2\, bh$

Area = $\sqrt{s(s-a)(s-b)(s-c)}$ where $s = (a + b + c)/2$

Parallelogram

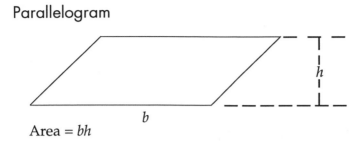

Area = *bh*

Trapezoid

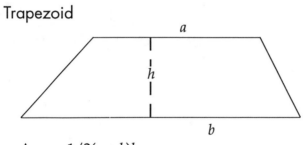

Area = 1/2(*a* + *b*)*h*

Circle

Circumference = $2\pi r$
Area = πr^2

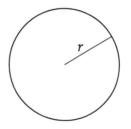

INDEX

Also from Sheridan House

Marine Electrical and Electronics Bible
by David Payne

How to select, install, maintain, and troubleshoot every electrical and electronic system on a boat.

Handbook of Offshore Cruising
by Jim Howard

Detailed, practical guidance on everything you need to know about long-distance cruising.

Marine Inboard Engines
by Loris Goring

Covers all the details about selecting, installing, and maintaining diesel and gasoline inboards.

Boat Handling Under Power
by John Mellor

A step-by-step guide to the principles and mechanics of handling powerboats and sailboats.

Surveying Small Craft, 3rd. ed.
by Ian Nicolson

The book to have when trying to detect problems in fiberglass, wood, and metal boats.

Using GPS
by Conrad Dixon

Hands-on advice for getting the most out of your GPS unit.

Boat Data Book, 3rd. ed.
by Ian Nicolson

Pure, indispensable data for anyone designing, building, fitting out, or repairing a boat.

America's Favorite Sailing Books